High-Speed System and Analog Input/Output Design

Thanh T. Tran

High-Speed System and Analog Input/Output Design

Second Edition

 Springer

Thanh T. Tran
Rice University
Houston, TX, USA

ISBN 978-3-031-04956-9   ISBN 978-3-031-04954-5   (eBook)
https://doi.org/10.1007/978-3-031-04954-5

This Springer imprint is published by the registered company Springer Nature Switzerland AG
The registered company address is: Gewerbestrasse 11, 6330 Cham, Switzerland

To my wife, Nga

Preface

This book is the second edition of the High-Speed DSP and Analog System Design. It covers the high-speed system and analog input/output design techniques and highlights common pitfalls causing noise and electromagnetic interference problems engineers have been facing for many years. The material in this book originated from my high-speed DSP system design guide (Texas Instruments SPRU 889), my system design courses at Rice University, and my experience in designing computers and DSP systems for more than 30 years. The book provides hands-on, practical advice for working engineers and electrical engineering students, including:

- Tips on cost-efficient design and system simulation that minimize late-stage redesign costs and product shipment delays.
- Fifteen easily-accessible chapters in 271 pages.
- Emphasis on good high-speed and analog design practices that minimize both component and system noise and ensure system design success, including industry compliance checks.
- Guidelines to be used throughout the design process to reduce noise and radiation and to avoid common pitfalls while improving quality and reliability.
- Hand-on design examples focusing on audio, video, analog filters, DDR memory, USB 3.1, and power supplies.

The inclusion of analog systems and related issues cannot be found in other high-speed design books.

This book is intended for practicing engineers and electrical engineering students and is organized as follows:

- **Chapter 1**: Highlights challenges in designing video, audio, computer, and communication systems.
- **Chapter 2**: Covers system design methodology, including pre-layout and post-layout design and simulations.
- **Chapter 3**: Reviews fundamentals of Alternate Current (AC) and Direct Current (DC).

- **Chapter 4**: Covers analog active and passive filter design including operational amplifier design with single-rail and dual-rail power supplies.
- **Chapter 5**: Presents an overview of data converter, sampling techniques and quantization noise.
- **Chapter 6**: Covers transmission line theories and effects. Demonstrates different signal termination schemes by performing signal integrity simulations and lab measurements.
- **Chapter 7**: Covers transmission line effects in frequency domain. Reviews Scattering Parameters or *S*-Parameters and shows how the models are being used in high-speed digital systems.
- **Chapter 8**: Shows the effects of crosstalk and methods to reduce interference. Highlights the importance of current return paths.
- **Chapter 9**: Provides memory sub-system design considerations, including DDR overview, signal integrity, and design example. And how to use the tool to do DDR compliance tests.
- **Chapter 10**: Provides a USB 3.1 design example using *S*-Parameter models and simulation tool to run compliance checks.
- **Chapter 11**: Covers design considerations of analog phase-locked loop (APLL) and digital phase-locked loop (DPLL) and how to isolate noise from affecting APLL and DPLL jitter.
- **Chapter 12**: Provides an overview of switching and linear power supplies and highlights the importance of having proper power sequencing schemes and power supply decoupling.
- **Chapter 13**: Covers the analytical and general power supply decoupling techniques, AC and DC resistance, and input/output filtering techniques.
- **Chapter 14**: Covers printed circuit board (PCB) stackup and signal routing considerations.
- **Chapter 15**: Describes sources of electromagnetic interference (EMI) and how to mitigate them.

Houston, TX, USA   Thanh T. Tran

Acknowledgments

I would like to thank many of my ex-colleagues at Texas Instruments Incorporated who encouraged me to write the original manuscript, High-Speed DSP and Analog System Design, and who helped me with doing many lab measurements to validate some of the theoretical concepts and simulations.

Also, I just can't thank Holly Jahangiri enough for taking time away from her retirement to review, edit, and provide invaluable suggestions. And, special thanks to Mary James and Pradheepa Vijay of Springer for giving me this writing opportunity and for providing outstanding support in completing all the logistics needed for the book. This book would not have been possible without the help and support from all these great individuals.

Finally, for my wife, Nga, I am still amazed with how she can hold a very demanding full-time job at HP and still find time to review the manuscript while continuously and unconditionally providing great care to the family.

Not to mention, I spent 9 years in graduate school while working full-time and was not home most of the time during those 9 years, she still found time to raise three kids, and all of them graduated from good universities (Massachusetts Institute of Technology, Trinity University, and St. Mary's University). She is truly amazing. With that said, I can't be happier than dedicating this entire book to her as part of my appreciation for everything she has done and for being a remarkable, long-life companion for more than 34 years.

Houston, TX, USA   Thanh T. Tran
2022

Contents

About the Author

Thanh T. Tran a professor of practice at Rice University teaching high-speed systems design and a server platform architect at Advanced Micro Devices (AMD), earned his B.S. in electrical engineering from the University of Illinois at Urbana-Champaign, and M.S. and Ph.D. in electrical engineering from the University of Houston. Before joining AMD, he was a senior principal SI/PI engineer at Raytheon Technologies, an engineering technologist at Dell Technologies and Compaq Computer, a chief technical advisor at Halliburton Company, and a CTO/senior engineering manager of New Emerging End Equipment in DSP Systems at Texas Instruments.

Dr. Tran, Senior IEEE member, has authored two books, has published over 24 technical papers, and currently holds 37 issued patents plus 9 patents pending related to virtual reality, computer gaming/audio systems, PC-based HDTV, oil and gas logging systems, fiber optic communication, high bitrate telemetry systems, and mixed analog/digital ASICs. The issued patents include two key inventions in high-speed systems design, AC-coupled sinewave clocking in multi-CPU servers and direct sequence spread spectrum clocking with adjustable EMI reduction.

Dr. Tran has received numerous prestigious awards, including two Achievement Awards and one 2021 Innovators Award from Raytheon, Bronze Excellent@Dell

Award, Texas Instruments Patent Hall of Fame, Halliburton MVP and Outstanding Contributions to Innovation and Technology Awards, Compaq Computer Cumulative Patent Award, and the first Entrepreneur/Innovation Award from the University of Houston.

Chapter 1
Challenges in High-Speed Systems Design

As system performance levels and clock frequencies continue to rise at a rapid rate, managing noise, radiation, and power consumption becomes an increasingly important issue. At high frequencies, the traces on a PCB carrying signals act as transmission lines and antennas that can generate signal reflections and radiations that cause distortion and create challenges in achieving timing requirements, industry interconnect protocol compliance, and electromagnetic compatibility (EMC) compliance. These can often make it difficult to meet Federal Communication Commission (FCC) Class A and Class B [1] requirements. Heat sinks and venting that may be required to address the thermal challenges of high-performance designs can further exacerbate EMC problems. Many systems today have embedded wireless local area network (WLAN) and Bluetooth, which will create further difficulties as intentional radiators are designed into the system.

With these difficulties, it is necessary to rethink the traditional high-speed system design process. In the traditional approach, engineers focus on the functional and performance aspects of the design. Noise and radiation are considered only toward the later stages of the design process if prototype testing reveals problems. But today, noise problems are becoming increasingly common and more than 70% of new designs fail first-time EMC testing. As a result, it is essential to begin addressing these issues from the very beginning of the design process. By investing a small amount of time in the use of low-noise and low-radiation design methods, and simulations at the beginning of the development cycle, this will generate a much more cost-efficient design by minimizing late-stage redesign costs and delays in the product ship date.

1.1 High-Speed Systems Overview

Typical systems such as the ones shown in Figs. 1.1 and 1.2 consist of many external to CPU or SoC devices such as Chipset, audio CODEC, video, LCD display, wireless communication (Bluetooth, GPS, UWB, and IEEE 802.11), Ethernet

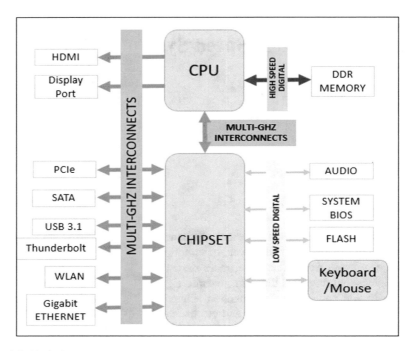

Fig. 1.1 Typical computer system

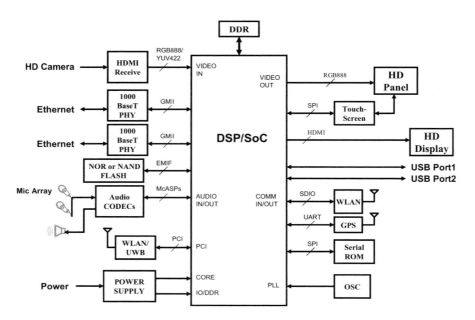

Fig. 1.2 Typical embedded system

Fig. 1.3 Typical noise path

controller, USB, power supply, oscillators, storage, memory, and other supporting circuitries. Each of these components can either be a noise generator or be affected by interferences generated by neighboring components. Therefore, applying good high-speed design practices and simulating high-speed ports are necessary to minimize both component and system-related noise and to ensure interconnect design compatibility. Multi-gigahertz serial interconnects are increasingly becoming more popular in high-speed systems for connecting one device to another as shown in Fig. 1.1. Designing these multi-gigahertz interconnects requires digital engineers to learn and apply techniques commonly being done by RF and microwave engineers, for example, scattering parameters or s-parameters.

The coupling between a noise source and noise victim causes electrical noise. Figure 1.3 shows a typical noise path. The noise source is typically a fast-switching signal, and the noise victim is the component carrying the signal. The noise victim's performance will be impacted by the noise. Coupling takes place through the parasitic capacitances and mutual inductances of the adjacent signals and circuits. Electromagnetic coupling occurs when the signal traces become effective antennas, which radiate and generate interferences to the adjacent circuitries.

There are many mechanisms by which noise can be generated in an electronic system. External and internal CPU/DSP clock circuits generally have the highest toggle rates and the primary source of high-frequency noise. Improperly terminated signal lines may generate reflections and signal distortions. Also, improper signal routing, grounding, and power supply decoupling may generate significant ground noise, crosstalk, and oscillations.

Noise can also be generated within semiconductors [2] themselves:

- *Thermal noise*: Also known as Johnson noise is present in all resistors and is caused by the random thermal motion of electrons. Thermal noise can be minimized in audio and video designs by keeping resistance as low as possible to improve the signal-to-noise ratio.
- *Shot noise*: Shot noise is caused by charges moving randomly across the gate in diodes and transistors. This noise is inversely proportional to the DC current flowing through the diode or transistor, so the higher DC operating current increases the signal-to-noise ratio. Shot noise can become an important factor when the DSP system includes many analog discrete devices on the signal paths, for example, discrete video and audio amplifiers.
- *Flicker noise*: Also known as 1/f noise is present in all active devices. It is caused by traps where charge barriers are captured and released randomly, causing

random current fluctuations. Flicker noise is a factor of semiconductor process technologies, so DSP system design cannot reduce it at the source, but the focus should be on mitigating its effects.

- *Burst noise and avalanche noise*: Burst noise, also known as popcorn noise, is caused by ion contamination. Avalanche noise is found in devices such as Zener diodes that operate in reverse breakdown mode. Both types of noise are again related to the semiconductor process technology rather than system design techniques.

Since government regulates the amount of electromagnetic energy that can be radiated, systems designers must also be concerned with the potential for radiating noise to the environment. The main sources of radiation are digital signals propagating on traces, current return loop areas, inadequate power supply filtering or decoupling, transmission line effects, and lack of return and ground planes. It is also important to note that at Gigahertz speeds, heat sinks, and enclosure resonances can amplify radiation.

Noise in high-speed systems cannot be eliminated but it can be minimized to ensure that it is not interfering with other circuits in the system. The three ways to reduce noise are suppressing it at the source, making the adjacent circuits insensitive to the noise, and eliminating the coupling channel. High-speed design practices can be applied to minimize both component and system-related noise and improve the probability of system design success. This book will address all three areas by providing guidelines that can be used from the very beginning of the design process through troubleshooting to reduce noise and radiation to acceptable levels. The noise-sensitive interface examples shown in this document are focused on audio, video, memory, and power supply. The performances of these systems are greatly affected by the surrounding circuitries and how these circuits interfaced to the CPU or DSP/SoC.

1.2 Challenges of Audio System

Audio systems represent one of the greatest challenges for high-speed DSP design. Relatively small levels of noise often have a noticeable impact on the performance of the finished product. In audio capture and playback, audio performance depends on the quality of the audio CODEC being used, the power supply noise, the audio circuit board layout, and the amount of crosstalk between the neighboring circuitries. Also, the sampling clock must be stable to prevent unwanted sounds such as pops and clicks during playback and capture. Figure 1.4 shows a typical signal chain of the DSP audio design. Most DSPs include a Multi-Channel Buffered Serial Port or McBSP [3] for interfacing with external audio CODECs. Although this is a proprietary interface, it is configurable to work with the industry standard I^2S audio CODECs.

Fig. 1.4 DSP audio system

All the blocks shown in Fig. 1.4 from the ADC to the Amp stage are very sensitive to noise so any interference coupled to any of the blocks will propagate and generate unwanted audible sounds. Common audio design problems include:

- Noise coupled to the microphone input: Mic input typically has a very high gain (+20 dB) so a small amount of noise can generate audible sounds.
- Not having an anti-aliasing filter at the audio inputs.
- Excessive distortion due to gain stage and amplitude mismatch.
- Excessive jitter on audio clocks, bit clock, and master clock.
- Lack of good decoupling and noise isolation techniques.
- Not using a linear regulator with high power supply rejection to isolate noise from the audio CODEC.
- Not having good decoupling capacitors on the reference voltage used for ADC and DAC converters.
- Switching power supply noise coupled to the audio circuits.
- High impedance audio traces are adjacent to noisy switching circuits and no shortest current return path is provided in the printed circuit board (PCB) layout to minimize the current return loop between the DSP and the CODEC.
- Not having isolated analog and digital grounds.

In summary, having good audio performance requires proper design of the ADCs, DACs, DSP interfaces, clocks, input/output filters, power supplies, and the output amplifier circuits. The performance of all these circuits not only depends on how well the circuits are being designed but also on how the grounds and power being isolated and the PCB traces being routed.

1.3   Challenges in Video System Design

Video processing is another important DSP application that is highly sensitive to noise and radiation. One of the major challenges of video systems design is how to eliminate video artifacts such as color distortion, 60 Hz hum, visible high-frequency interferences caused fast-switching buses, audio beat, etc. These issues are generally related to improper video board design and PCB layout. For example, power supply noise may propagate to the video DAC output, audio playback may cause transients in the power supply, and the high-frequency radiations may couple back to the tuner. Here are some common video noise issues:

- Signal integrity, excessive overshoots, and undershoots on the HSYNC, VSYNC, and pixel clocks caused by improper signal terminations.
- Excessive radiations from high-speed buses such as PCI, parallel video ports (BT.1120, BT.656), and DDR memory.
- Excessive encoder, decoder, and pixel clock jitter cause problems with detecting the color information. For example, the color screen only displays black and white images.
- The lack of video termination resistors will cause distortion of the video image. A 75-Ω termination resistor must be used at the input of the video decoder and the output of the video encoder.
- Audio playback may cause a flicker on the video screen. This can be corrected by increasing the isolation of the video and audio circuits. The best method is by using high power supply rejection ratio (PSRR) linear regulators to isolate the audio CODEC and the video encoder/decoder supplies. Also, manually route the critical traces away from any of the switching signals to reduce the crosstalk and interferences.
- An isolated analog ground without a high-speed signal return path. It is important to remember that for a low-speed signal, below 10 MHz, current returns on the lowest resistance, which usually is the shortest path. High-speed current, on the other hand, returns on the path with the least inductance, usually underneath the signal.

Figure 1.5 shows a typical DSP HD video system where the analog video signals, high definition (HD) and standard definition (SD), are captured, processed, and then displayed. The quality of this video signal path determines the video performance of the display, especially at the input video stages and at the output video stages. Since the system design and layout are critical, it is necessary to apply the high-speed design rules discussed in this book to reduce the negative effects of the switching noise, crosstalk, and power supply transients in order to reduce or eliminate video artifacts. In this system, the digital video inputs and outputs, such as high-definition

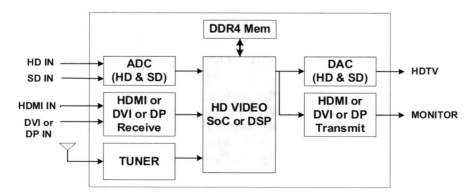

Fig. 1.5 DSP HD video system

media interface (HDMI), digital video interface (DVI), and DisplayPort (DP), are also highly sensitive to system noise as noise causes jitter which increases the bit error rate (BER). As in any electronic system, it is not possible to eliminate the noise totally but applying good design techniques will reduce the risk of it having a negative impact on performance.

In any HD video system, there are many wide high-speed buses switching at a rate of 66 MHz or higher, and these buses generate broadband noise and harmonics that cause radiations in the Gigahertz range. This type of interference is difficult to control because there are so many of these busses on the board and it is not practical to terminate every signal trace being routed from one point to another. The good news is that there are good design practices to follow in order to minimize interference.

1.4 Challenges in Communication System Design

Like video and audio systems, communication is another important DSP application that is highly sensitive to noise and radiation. One of many challenges here is creating systems with multiple powerful and highly integrated DSPs that deliver high performance with very low bit error rate and interference. In these systems, interference not only generates EMI problems but also jams other communication channels and causes false channel detection. These issues can be minimized by applying proper board design techniques, shielding, RF, and mixed analog/digital signals isolation. In some cases, a spread spectrum clock generator may be required to further reduce the interference and to improve the signal-to-noise ratio. Although spread spectrum clock reduces the peak level radiation, the harmonics of this clock are spreading over a wider bandwidth, and this can cause inter-channel interference so engineers must be careful when using this type of clock generator circuit. Table 1.1 shows high-speed buses generating harmonics that interfering with embedded Wireless Local Area Networks (WLAN). One example of the communication

Table 1.1 High-speed busses interference [4]

Standards	Wireless networking	Interfering clocks and busses
Bluetooth	Personal area network (2.4 GHz band)	Gigahertz Ethernet, PCI Express, Display clock harmonics
IEEE 602.11b/g	WLAN (2.4 GHz band)	Gigahertz Ethernet, PCI Express, Display clock harmonics
IEEE 802.11n	High-speed WLAN (5 GHz band)	Gigahertz Ethernet, PCI Express, Display clock harmonics
IEEE 802.16e	Mobile broadband (Wi-Max, 10–66 GHz band)	PCI Express, Display harmonics
IEEE 802.11a	WLAN (5 GHz band)	PCI Express, Display clock harmonics

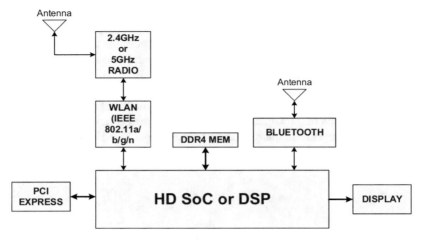

Fig. 1.6 DSP communication system

systems is shown in Fig. 1.6 where both Bluetooth and IEEE 802.11 are being implemented on the same motherboard and residing on the same 2.4 GHz RF spectrum. The most difficult tasks are how to prevent the two systems from interfering with each other and how to prevent radiations from the high-speed busses (PCI Express, DDR4, and display) interfering with the embedded antennas. By applying the rules outlined in this book, engineers will improve and increase the probability of design success.

1.5 Challenges of Computer System Design

Modern computer system consists of graphics, video, audio, communication, and peripherals (keyboard, mouse, monitor). The computer design is becoming more like multi-gigahertz Multiple Inputs Multiple Outputs (MIMO) design as many high-speed ports are running as high as 32 Gbps rate as shown in Fig. 1.7.

The challenges engineers are facing in designing computer systems today are:

- Applying RF/microwave techniques such as s-parameters modeling and simulation to developing multi-gigahertz ports.
- 3D electromagnetic (3D EM) modeling of the channels and vias must be done to guarantee timings.
- Industry standards such as DDR Memory, USB 3.1, PCIe, or 10 GigE compliance tests must be completed before releasing the layout to fabrication.
- Completing signal integrity and power integrity, including step responses in time domain and target impedances in frequency domain.

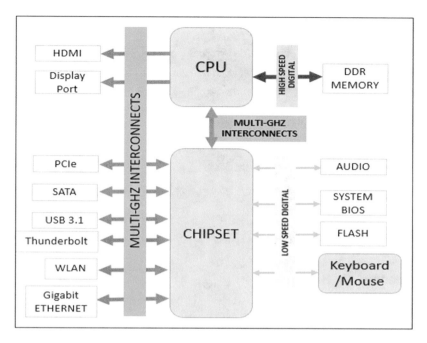

Fig. 1.7 Modern computer system

- Evaluating electromagnetic radiation (EMI) and RF immunity. Engineers not only have to mitigate radiations but need to harden the design to improve RF immunity. Interference within a system increases with operating frequency and could cause system failures.

1.6   Summary

This chapter highlights many challenges engineers face today. The good news is that many of the issues can be mitigated by applying good design practices and using state-of-the-art tools to simulate the circuits as described in this book.

The chapter arrangement of the book follows the input to output signal flow as shown in Fig. 1.8 where a number in the circle denotes the chapter number of the book, and the orange paths are high-speed digital signal paths while black paths are analog input/output signals. On the soft copy of the book, clicking on the circle advances to the chapter labeled in the circle.

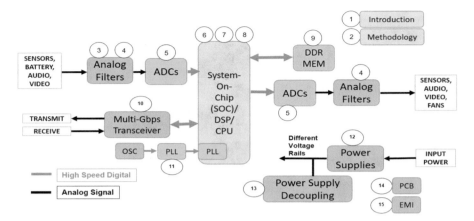

Fig. 1.8 Analog/digital signal flows of a system

References

1. Federal Communication Commission, Unintentional Radiators Title 47 (47CFR), Part 15 B (2005). http://www.fcc.gov/oet/info/rules/part15/part15-91905.pdf
2. T. Hiers, R. Ma, TMS320C6000 McBSP: I^{2}S Interface. Texas Instruments Inc's Application Report, SPRA595 (1999)
3. S. Franco, *Design with Operational Amplifiers and Analog Integrated Circuits* (McGraw-Hill, New York, 2002)
4. M. Nassar, K. Gulati, M. DeYoung, B. Evans, K. Tinsley, Mitigating near-field interference in laptop embedded wireless transceivers. IEEE J. Signal Process. Syst. **63**, 1–12 (2008)

Chapter 2
System Design Methodology

In addition to applying good design practices, designers must go through and thoroughly analyze the design, including PCB parasitics, to improve the probability of success after PCB fabrication. The proposed analysis is based on using multiple tools to perform different types of simulations of the design as there is no one tool that can do all the analysis accurately.

This chapter covers high-speed system design methodology and how to leverage different simulation tools (HSPICE [1], TINA [2], HYPERLYNX [3]) to do analysis.

2.1 System Design Methodology

The flowchart below shows a methodology of taking a design from schematics to layout (Fig. 2.1).

2.2 Pre-layout Analysis

In this step, it is critical to define the optimized floorplan of the layout. Separating analog and digital sections is a must, including using high-speed and low-speed current loop methods to determine the best placement of the components.

To prepare for the PCB layout, here is a list of recommended tasks.

- Component placement—separating analog and digital components and placing the components in such a way that minimizes the routing lengths.
- Determining high-speed signal paths, especially signals running at 1 Gbps or higher. A good rule is to keep 2 vias per trace maximum, and the via must be

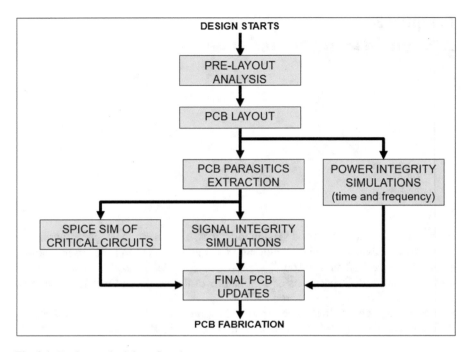

Fig. 2.1 Design methodology flowchart

placed as close to the source or load as possible. It is recommended to add a ground via next to the signal via if the signal is transitioned from one layer to another. This allows a minimum current return loop.

- Deciding microstrip or stripline traces, and tuning characteristic impedance to match the intended targets.
- Simulating critical traces and circuits. Using HyperLynx LineSim to emulate PCB.

2.3   PCB Layout

A routing document should be created for layout engineer to follow. This document includes special routing guides generated by system designers.

2.3.1   PCB Parasitics Extraction

At the completion of the PCB layout, a post-layout analysis starts by extracting PCB parasitics to use in the simulations to determine if the layout causes any changes in the circuit or system design performance. Tuning the circuit after layout typically is required to meet the design targets because PCB parasitics are almost always affecting high-speed traces.

A tool such as Mentor Graphics HyperLynx can extract parasitics accurately and generate an output in spice format or s-parameter models to use in the simulator. A detailed explanation of s-parameters will be covered in the later chapter of this book.

Example 2.1 PCB Extraction The layout in Fig. 2.2 only contains four 50-Ω transmission lines with and without vias. The file is imported to HyperLynx, and HyperLynx generates spice and s-parameter outputs as shown below (experimental layout).

The picture in Fig. 2.3 shows TLine2 being selected for extraction and the spice file of the transmission line and the via. This spice file can be incorporated into a system for simulating circuits with parasitics.

The picture in Fig. 2.4 shows an s-parameter model created by HyperLynx, and this model can be included in the system to simulate the design with parasitics. s-Parameter is commonly being used as this is an industry standard, Touchstone format, which is supported by many spice-based tools. Details of s-parameter model will be covered in the later chapter.

2.4 Spice Simulation of Critical Circuits

The performance of critical circuits like RF filters and high-speed SERDES channels is often affected by PCB layout, so it is particularly important for designers to extract parasitics and include them in the system simulation. One example is to use HSPICE to do critical timing analysis or to plot frequency response of an RF filter.

Figure 2.5 shows a proposed methodology for designing an RF filter.

Figure 2.6 is a circuit synthesized by free online software, Marki Microwave [4]. The circuit is a fifth-order bandpass filter with cutoff frequencies at 700 MHz and 1.1 GHz.

The picture below shows a TINA (TI Analog Circuit Simulator) simulation of the bandpass filter. This confirms the corner frequencies of the filter without any PCB parasitics (Fig. 2.7).

Fig. 2.2 Experimental layout

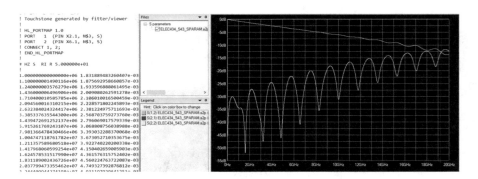

```
W_p_T5 N=1 via_4_000000e_03_1_250000e_03_layer_1 0 X6_1 0
+ RLGCmodel=p_T5_model L=4.8234600e-02 FP=1.0000000e+09 MULTIDEBYE=1
+ Crgh=6.3327533e+00 Frgh=5.0691062e+09
.MODEL p_T5_model W MODELTYPE=RLGC N=1
+ Lo =
+   2.7016924e-07
+ Co =
+   1.1014600e-10
+ Ro =
+   6.0340178e-01
+ Go =
+   0.0000000e+00
+ Rs =
+   6.6576009e-04
+ Gd =
+   1.1238963e-11

X_VIA via_4_000000e_03_1_250000e_03_layer_1 via_4_000000e_03_1_250000e_03_layer_4 VIA
X_VIA_1 via_2_000000e_03_1_250000e_03_layer_1 via_2_000000e_03_1_250000e_03_layer_4 VIA_1

******************************
* Define subcircuit VIA
.subckt VIA l_1 l_4
C_p_C1 0 l_1 4.4631366e-14
C_p_C2 0 p_2 2.4435171e-14
C_p_C3 0 p_4 7.5844716e-14
C_p_C4 0 l_4 8.2964286e-14
L_p_L1 p_1 l_1 2.3930832e-11
L_p_L2 p_3 p_2 2.3822045e-12
L_p_L3 p_5 p_4 6.5218066e-10
R_p_R1 p_2 p_1 7.5936158e-05
R_p_R2 p_4 p_3 1.0404387e-05
R_p_R3 l_4 p_5 6.1902702e-04
.ends VIA
**********************
```

Fig. 2.3 Example of spice extraction of TLine2

```
! Touchstone generated by fitter/viewer
!
! HL_PORTMAP 1.0
! PORT    1  (PIN X2.1, N$3, S)
! PORT    2  (PIN X6.1, N$3, S)
! CONNECT 1, 2;
! END_HL_PORTMAP
!
# HZ S  RI R 5.000000e+01
!
1.00000000000000e+06 1.83189483260407e-03
1.10000001490116e+06 1.87569295860857e-03
1.24000003576279e+06 1.93359688061495e-03
1.43600000649696e+06 2.00908026259127e-03
1.71040001058578e+06 2.10601001650045e-03
2.09456001631021e+06 2.22857180224589e-03
2.63238402432441e+06 2.38122497571169e-03
3.38533763554430e+06 2.56870375927376e-03
4.43947269125213e+06 2.79606901757933e-03
5.91526176924310e+06 3.06880875603898e-03
7.98136647843046e+06 3.39303228837006e-03
1.00474711876178e+07 3.67305271035367e-03
1.21135758968051e+07 3.92274022020033e-03
1.41796860599254e+07 4.15040265900500e-03
1.62457853151779e+07 4.36157631575240e-03
1.83118900243672e+07 4.56022476372240e-03
2.03779947335546e+07 4.74932739287681e-03
```

Fig. 2.4 s-Parameter simulation

Now, add PCB parasitics (extracted s-parameter model) to the bandpass filter and do an HSPICE simulation to check the filter's performance. The simulation results with parasitics as shown in the picture below show that the corner frequencies of the filter shift from 700 to 520 MHz on the low end and from 1.1 GHz to 750 MHz on

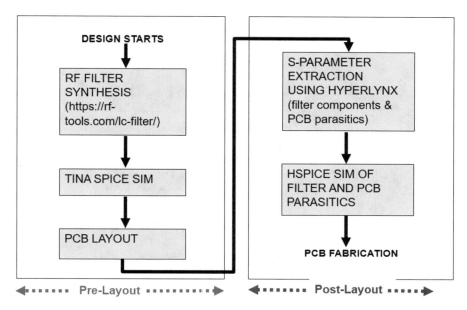

Fig. 2.5 Flow of spice simulation of extracted PCB

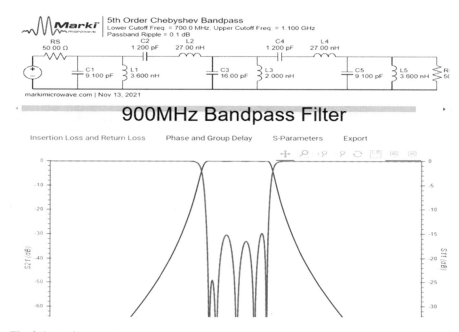

Fig. 2.6 RF filter synthesis

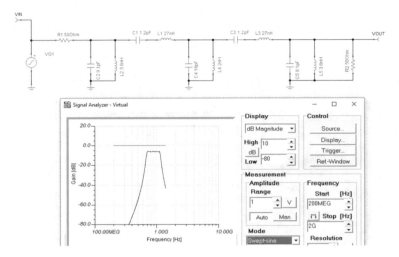

Fig. 2.7 TINA simulation of the filter

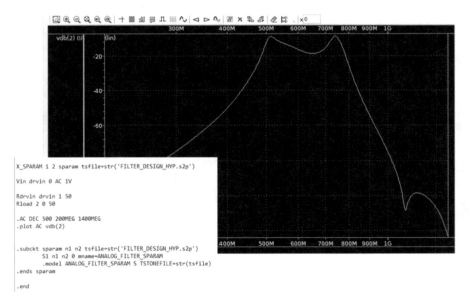

Fig. 2.8 HSPICE simulation of filter with parasitics

the high end. This demonstrates that the PCB parasitics can significantly change the performance of the filter. In this case, tuning is required to compensate for the PCB layout and to achieve the filter specifications (Fig. 2.8).

The Power Integrity and Signal Integrity tasks will be covered in detail in the later chapters.

2.5   Summary

It is important for designers to use the right tool for the task. For example, HyperLynx is a useful tool for PCB extractions and traditional digital signal integrity simulations, while HSPICE is an excellent tool for critical timing analysis, including simulating s-parameters. Another excellent tool for extracting layout is ANSYS SiWave.

In summary, designers should develop their own methodology using the tools available for them, and here is a list of tasks to do.

Perform circuit or logic simulations to validate functionalities of the design before doing the actual PCB layout.

Complete floorplanning, including analog and digital components placement and separation.

Do pre-layout simulations to determine critical paths, maximum number of vias allowed, and routing guidelines.

For post-layout simulations, be sure to check critical circuits to make sure to compensate for parasitics if necessary.

References

1. Synopsys HSPICE Spice Circuit Simulator
2. TINA-TI SPICE-Based Analog Simulation Program. https://www.ti.com/tool/TINA-TI
3. Mentor Graphics (HyperLynx VX2.7) HyperLynx Signal Integrity Simulation software. http://www.mentor.com/products/pcb-system-design/circuit-simulation/hyperlynx-signal-integrity/
4. Marki Microwave, Inc., RF Tools. https://rf-tools.com/lc-filter/

Chapter 3
AC Versus DC

One of the most important concepts in electrical engineering is being able to do AC and DC analysis of a circuit or system. All circuits or systems must have DC voltages to stay active and to process AC signals while AC signals are signals of interest, which typically have a single or multiple frequency components superimposed on an AC waveform. For example, squarewave is a summation of multiple frequency sinewaves.

This chapter covers the fundamentals of AC-coupled and DC-coupled signals and how these signals propagate from the inputs to the outputs of a system.

3.1 Alternating Current (AC) and Direct Current (DC)

The flow of electric charge that periodically reverses direction is referred to as alternating current, while the flow of electric charge in direct current only flows in one direction. In Fig. 3.1, $Vac1$ is an AC voltage source with a frequency greater than zero, and $V1$ is a 5 V DC bias voltage with a frequency of zero. At steady-state, $C2$ acts as an open circuit for DC because the impedance of $C2$ goes to infinity when the frequency is zero as in Eq. (3.1).

$$Zc2 = \frac{1}{2\pi f C2},\tag{3.1}$$

where $Zc2$ is the impedance of $C2$, and f is the frequency.

In this circuit, since there is no DC current flowing through $C2$, the voltage drop across resistors $R3$ and $R4$ is zero. This makes $Vout1$ potential the same as $V1$, 5 V.

Now, if an AC signal is applied to $Vac1$, this signal is level-shifted to the bias voltage, $V1$. And the amplitude of this AC signal equals to

T. T. Tran, *High-Speed System and Analog Input/Output Design*,
https://doi.org/10.1007/978-3-031-04954-5_3

$$\text{Amplitude} = \frac{R3}{R3 + R4} Vac1, \tag{3.2}$$

assuming $C2$ is very large and the impedance of $C2$ is very small at all frequencies as compared to $R3$ and $R4$. For AC signals, $V1$ 5 V source acts as an AC ground.

Example 3.1 Using TI TINA Analog Simulator [1]
For $C2 = 100 \, \mu F$, $R3 = 1 \, k\Omega$, and $R4 = 1 \, k\Omega$, $V1 = 5$ VDC and $Vac1 = 2$ V peak-to-peak at 1 kHz,

$$Zc2 = \frac{1}{2\pi fC} = \frac{1}{2\pi(1000)(100e - 6)} = 1.6 \, \Omega,$$

which is small as compared to $R3$ and $R4$ and is negligible.

DC component of $Vout1$ equals 5 V, and AC signal is

$$Vout1 = \frac{R3}{R3 + R4} Vac1 = \frac{1K}{1K + 1K}(2Vpp) = 1Vpp.$$

Waveforms of spice simulations of circuit in Fig. 3.1 are shown in Fig. 3.2. As shown in the calculations, the output voltage is $1Vpp$ with a DC bias voltage of 5 V.
Rules to Remember

- DC sources are AC ground. Connect DC sources to ground when doing AC analysis.
- AC signals are signals of interest, while DC voltages are generally for biasing.
- Always keep DC sources as stable as possible, power supply decoupling capacitors are needed.
- AC-coupled signals have no DC component unless they are biased again on the other side of the AC-coupling capacitor.

The trend in the industry is moving toward serial busses, incorporating AC-coupled methods to isolate DC voltages. This way DC voltage of one circuit

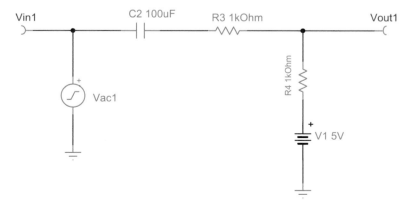

Fig. 3.1 AC-coupled circuit

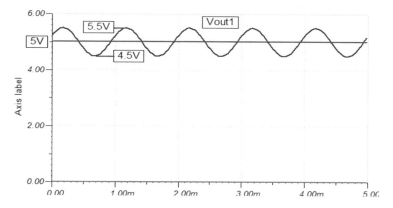

Fig. 3.2 DC bias of an AC waveform

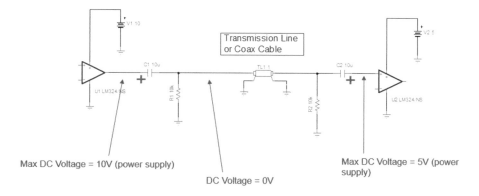

The plus side of the capacitor must be on the side with higher DC voltage

Fig. 3.3 AC-coupled circuit

does not affect another circuit. Some of the AC-coupled methods used in many industry standards are HDMI, DisplayPort, PCIe, Analog Audio Inputs/Outputs. These busses enable users to connect devices together and start operating with almost no user interactions required. In some cases, a system may require users to enter security information before enabling access.

These ease-of-use interconnections are possible because of intelligent software being able to identify devices and of solid hardware design that allows circuits to interact. Software identification methods are beyond the scope of this book. For hardware, the key enabler is to isolate DC voltages from one system to another by using AC-coupling technique discussed in the previous section.

In Fig. 3.3, the AC-coupling capacitors in series with the transmit and receive channels prevent DC voltages of one system affecting the other, and only allow AC signals to propagate. Without the AC-coupling capacitors, the transmitted signals are

Fig. 3.4 AC-coupled audio circuit

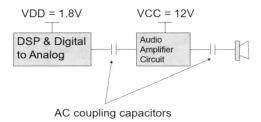

AC coupling capacitors

Fig. 3.5 AC-coupled digital logic

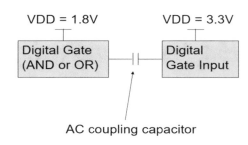

AC coupling capacitor

shifted to a DC biased voltage to ensure signals have a maximum symmetrical swing at the transceiver output, but the receiver input operates at a different DC voltage, and this causes DC mismatch between the transceiver and the receiver. This mismatch in DC voltages leads to excessive signal distortion.

Having AC-coupling capacitors allows the transmitter and receiver to independently generate their own bias voltages without impacting each other's performance. The only requirement here is not to overdrive the receiver by transmitting an AC voltage that is higher than the maximum allowable limit. The key is if using polarized capacitors, such as electrolytic or tantalum, the positive side must be connected to the higher DC voltage side of transmitter or receiver as shown in Fig. 3.3.

This AC-coupling scheme is being used in many different interconnect interfaces in the industry. Some of the examples are:

- PCIe for option cards, gigabit networks, 3D graphics accelerators.
- Video equipment, HDMI, analog camera cables.
- Audio equipment, audio amplifiers, audio headsets.

Example 3.2 AC-Coupling Design
1. Analog circuits with different power supply voltages (Fig. 3.4).
2. Standard digital logic cannot be AC-coupled. For digital design, a voltage translator must be used to interface two logic blocks running different power supply voltages. The capacitor in Fig. 3.5 needs to be replaced with a voltage translator for digital logic to work properly. This is because the capacitor formed an RC circuit with the load resistance acts as a differentiator which generates voltage spikes based on the edge rates of the digital signals and severely changes the shape of the digital waveforms.

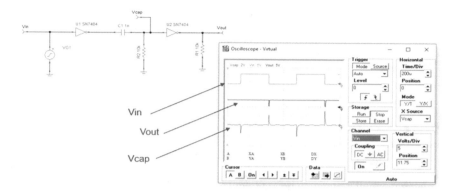

Fig. 3.6 AC-coupled digital logic simulation

Figure 3.6 shows why having an AC-coupling capacitor placed in the middle of the two inverters does not function properly. The RC circuit in time domain digital logic acts as a differentiator which differentiates the input waveform and generates positive and negative spikes. The amplitude and width of spikes shown in Fig. 3.6 depend on the rise and fall times of the digital signal. This causes a logic failure because the output of the second inverter is not equal to the input signal when the input signal is inverted twice.

3.2   Avoiding Pitfalls in AC-Coupled and DC-Couple Circuits

As described in the previous section, an AC-coupling capacitor is required to isolate DC voltages of two cascading circuits running on different power supplies. Here are the common pitfalls and how to avoid them.

- AC-coupling capacitor can generate a large noise spike during powering up and down of the drive circuit. In Fig. 3.7 circuit, the audio amplifier output is set at half of the power supply voltage to guarantee the maximum symmetrical swing at the output. During powering up, the amplifier output goes from 0 to 6 V (half of the 12 V supply) quickly, and this transient generates a large spike at the speaker output which can cause audible clicks or pops at the speaker, depending on the value of the capacitor and the slewrate of the DC voltage going from 0 to 6 V at the speaker output.
- There are ways to mitigate this issue:

Fig. 3.7 AC-coupled
circuit

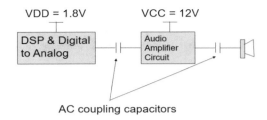

Fig. 3.8 Bridge-tied
speaker load

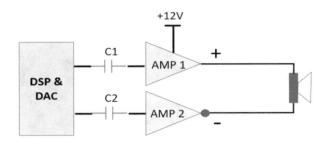

- Mute the amplifier output by using a circuit that is not tied to power supply
  ramping up and down. For example, incorporating a similar circuit in [2].
- Implement a bridge-tied load as shown in Fig. 3.8. In this case, no
  AC-coupling capacitor is needed because both inverting and non-inverting
  outputs of amplifier are equal, no DC current flowing through the speaker to
  generate noise spikes.

- Using a wrong AC-coupling capacitor value is one of the common pitfalls. To
  figure the minimum value of the AC-coupling capacitor, it is necessary to find the
  input impedance of the circuit being driven by the previous stage and the
  minimum operating frequency.

  In Fig. 3.8 example, to find the capacitor $C1$ value for the AMP 1 input
  impedance of 20 kΩ and the minimum operating frequency of 20 Hz, here is
  the calculation.

$$C1 = \frac{1}{2\pi f Z\text{in}} = \frac{1}{2\pi(20)(20e+3)} = 0.4 \ \mu F,$$

  where f is the -3 dB corner of the operating frequency and Zin in the input
  impedance of AMP 1. More filter design details will be covered in Chap. 4.
- For DC-coupled circuit, such as the bridge-tied load in Fig. 3.8, the importance
  here is to control the two DC ramp rates at the outputs of the inverting and
  non-inverting amplifiers to be synchronous so that there is no DC current flowing
  through the speaker during powering up and down, and normal operation.
- For digital designs, DC-coupled method is almost always being implemented. If
  the power supplies of the two cascading circuits are different, then the solution is
  to add a voltage translator to regenerate compatible voltage levels for the next

Fig. 3.9 DC-coupled
circuit

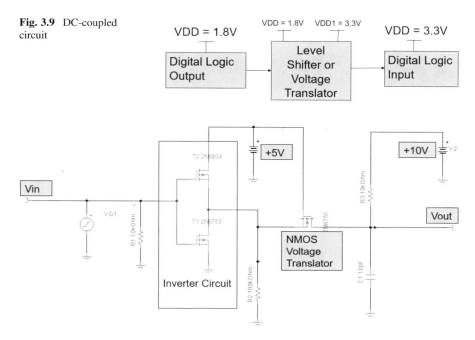

Fig. 3.10 TINA circuit of using NMOS voltage translator

stage as shown in Fig. 3.9. Voltage Translators are widely available, and one
low-cost example of designing a low-speed Voltage Translator is using one
N-Channel MOS transistor, $T3$, as in Fig. 3.10 where the input, Vin, is switching
from 0 to 5 V, and the output, $Vout$, is going from 0 to 10 V.

References

1. Texas Instruments TINA Analog Circuit Simulator
2. T. Tran, *Apparatus for Eliminating Audio Noise When Power is Cycled to a Computer*, Patent #
   5,768,601, Jun 1998

Chapter 4
Analog Filter Design

This chapter presents passive and active filter topologies and design techniques, including practical design examples and system simulations. In DSP systems, there are analog filters required for signal conditioning and limiting the bandwidth before sampling. To design these filters, designers need to be knowledgeable about operational amplifiers, DC biasing circuits, AC-coupling techniques, and traditional passive components like inductors, capacitors, and resistors.

4.1 Anti-Aliasing Filters

There are many types of anti-aliasing filters (Butterworth, Chebyshev, Inverse Chebyshev, Cauer, and Besser-Thomson) [1], but for video, audio, and communication applications, it is best to use the Butterworth filter because it has the best passband performance; it is known as a maximally flat passband filter. A Butterworth filter can be implemented in two ways: (1) With passive components such as resistors, inductors, and capacitors. (2) Active components such as operational amplifiers, resistors, and capacitors. There are advantages and disadvantages between the two approaches, so selecting the right topology requires designers to understand the basic differences and how one filter works better than the other. One important rule-of-thumb is the higher sampling frequency, the lower order anti-aliasing filter needed as shown in Chap. 7.

4.1.1 Passive and Active Filters Characteristics

Table 4.1 shows the characteristic of passive and active filters and highlights the advantages and disadvantages of each type.

© The Author(s), under exclusive license to Springer Nature Switzerland AG 2023
T. T. Tran, *High-Speed System and Analog Input/Output Design*,
https://doi.org/10.1007/978-3-031-04954-5_4

Table 4.1 Passive and active filters

Electrical characteristic	Passive filter	Active filter
Power supply	Not required	Required
Voltage gain	No	Yes
Filter components	Inductors, capacitors, resistors	Op amps, resistors, capacitors
Radiation	Yes, inductors tend to radiate	No, proper decoupling is required
Circuit stability	Very good	Good, proper analysis is required
Dynamic range limitation	No	Yes (power supply and op amp)
Depending on source and load impedance	Yes (filter characteristics are affected by the source and the load)	No (op amp has very high impedance inputs)
Reliability	Very good (passive components are very reliable)	Good (active circuits like op amps are not as reliable as passive components)
Frequency range	Very good	Good (limited by the op amp bandwidth)
Cost	More expensive (inductors)	Less expensive

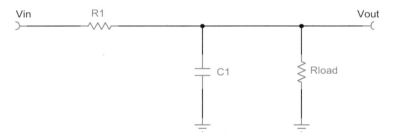

Fig. 4.1 First-order passive filter

In summary, if gain is not required and source and load impedances are known, then it is better to go with passive filters. Now, if impedance isolation and gain are required, then active filters would be better.

4.1.2   Passive Filter Design

4.1.2.1   First-Order Passive Lowpass Filter

The first-order passive filter can easily be realized by one resistor and one capacitor as shown in Fig. 4.1.

Assuming R_{load} is much higher than R_1, the -3 dB corner frequency for the filter in Fig. 4.1 is

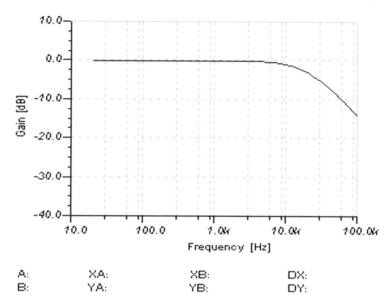

Fig. 4.2 First-order passive filter simulation

$$f_{-3\text{dB}} = \frac{1}{2\pi R_1 C_1}. \tag{4.1}$$

Example 4.1 An audio ADC has a bandwidth of 20 Hz–20 kHz and requires a first-order anti-aliasing filter at its input. Design and simulate [2] this filter.

Answer: Let $f_{-3\text{dB}} = 20$ kHz,

$$20\ \text{kHz} = \frac{1}{2\pi R_1 C_1}.$$

Now, let $C_1 = 0.001$ μF and solve for R_1. $R_1 = 8\text{K}\Omega$. As shown in Fig. 4.2, the −3 dB corner frequency is at 20 kHz.

4.1.2.2 Second-Order Passive Filter Design

The second-order lowpass circuit requires one inductor and one capacitor. In general, the order of the filter circuit is equal to the number of capacitors and inductors in the circuit. As mentioned in the previous section, the passive filter depends on the source and load impedances, so let us assume that the source impedance, R_S, is much higher than the load impedance, R_L, as shown in Fig. 4.3.

The frequency response of the second-order filter circuit has amplitude peaking at the −3 dB corner and the amount of this peaking depends on the ratio of R_L and R_S.

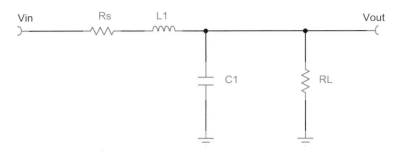

Fig. 4.3 Second-order passive filter circuit

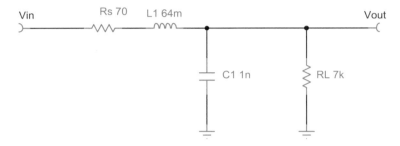

Fig. 4.4 Second-order passive filter circuit design

This peaking typically does not affect the circuit performance as long as the noise at the corner frequency is very low, so it is crucial for designers to verify that the frequency response and signal-to-noise over the passband are as expected. It is common to fine-tune R_L and R_S to get the frequency response needed for the application.

For $R_L \gg R_S$, the -3 dB corner frequency for the circuit in Fig. 4.3 is

$$f_{-3dB} = \frac{\sqrt{R_L + R_S}}{2\pi\sqrt{L_1 C_1 R_L}}. \tag{4.2}$$

Example 4.2 An audio ADC has a bandwidth of 20 Hz–20 kHz and requires a second-order anti-aliasing filter at its input. Design and simulate [2] this filter.

Answer: Let $f_{-3dB} = 20$ kHz, $R_S = 70\ \Omega$ and $R_L = 7K\ \Omega$.

$$20\ \text{kHz} = \frac{\sqrt{7K + 70}}{2\pi\sqrt{(7K)L_1 C_1}}.$$

Now, solving for L_1 and C_1

$L_1 C_1 = 6.388e^{-11}$, let $C_1 = 0.001$ μF, $L_1 = 0.064$ H. Figure 4.4 shows the final circuit diagram and Fig. 4.5 shows the simulation.

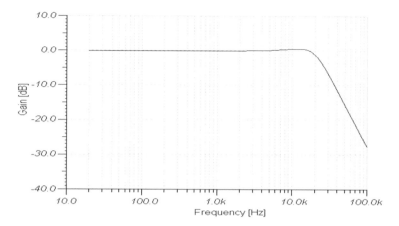

Fig. 4.5 Second-order passive filter simulation

As shown in Fig. 4.5, the −3 dB corner is approximately at 20 kHz. The amplitude peaking at this corner is determined by the source and load resistors. This is a major disadvantage of a second-order passive filter since it is not practical to control the load impedance. The input impedance of an ADC or load impedance varies greatly with the silicon process and the range is not specified in the datasheet. The datasheet typically only shows a minimum input impedance specification. One method to minimize the effect of the input impedance is selecting an ADC with high impedance and adding an external load resistor with a resistance much lower than the ADC input impedance so that the parallel combination of the two resistors will not vary widely over the silicon process.

4.1.3   Active Filter Design

For practical purposes, active filter designs in this section are focused only on first- and second-order Butterworth, as these are the most popular topologies being implemented in audio, video, and communication systems today. To be able to design active filters, designers need to understand operational amplifier or op amp and how to bias the op amp to get the maximum dynamic range.

4.1.3.1   Operational Amplifier (Op Amp) Fundamentals

An op amp circuit has three terminals, inverting input and non-inverting input and output, as shown in Fig. 4.6 and has the following electrical characteristics:

- Very high impedance: For an ideal op amp, it is assumed infinite input imped- ance. The input currents are very small and are negligible.

Fig. 4.6 Operational
amplifier

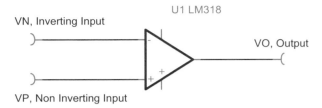

- Very low output impedance: It is zero for an ideal op amp.
- Virtual ground: The voltages at the negative and positive inputs are equal.
- Very high open loop gain: It is infinite for an ideal op amp.

The relationships between the input and output of the op amp are

$$V_D = V_P - V_N,\qquad(4.3)$$

where V_D is the differential input, V_P is the non-inverting input and V_N is the inverting input.

$$V_O = aV_D,\qquad(4.4)$$

where a is the open loop gain.
Substitute Eq. (4.3) into Eq. (4.4),

$$V_O = a(V_P - V_N).\qquad(4.5)$$

As shown in Eq. (4.5), the open loop op amp works like a comparator where the output voltage is the difference of the input multiplied by the open loop gain.

Op amp typically being implemented in a linear circuit such as filters works in a closed loop system where the output has a feedback to the input to control the signal or AC gain. In this case, the inverting and non-inverting inputs are equal; this is also known as virtual ground. The gain equations are derived as follows.

Non-Inverting Amplifier

Applying Kirchhoff's Current Law at node V_N,

$$I_2 = I_N + I_1,\qquad(4.6)$$

where I_N is the negative input current and is equal to zero since op amp has infinite input impedance (Fig. 4.7).

Fig. 4.7 Non-inverting
amplifier

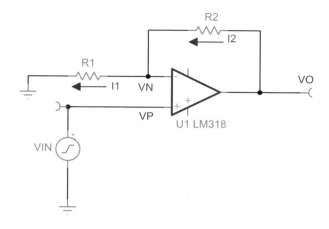

$$I_2 = \frac{V_O - V_N}{R_2},$$

$$I_1 = \frac{V_N}{R_1},$$

Substitute I_2 and I_1 into Eq. (4.6),

$$\frac{V_O - V_N}{R_2} = \frac{V_N}{R_1}. \tag{4.7}$$

Due to virtual ground rule,
$V_N = V_P$ and $V_P = V_{IN}$ as shown in the circuit.
Substitute V_{IN} into Eq. (4.7) and solve for the gain, V_O/V_{IN}.

$$\text{Gain} = \frac{V_O}{V_{IN}} = \frac{R_2}{R_1} + 1. \tag{4.8}$$

Inverting Amplifier

Applying Kirchhoff's Current Law at node V_N,

$$I_2 + I_1 = I_N, \tag{4.9}$$

where I_N is the negative input current and is equal to zero since op amp has infinite
input impedance (Fig. 4.8).

$$I_2 = \frac{V_O - V_N}{R_2},$$

$$I_1 = \frac{V_{IN} - V_N}{R_1},$$

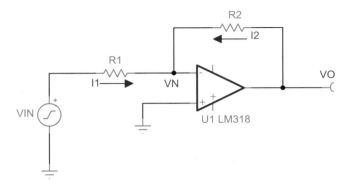

Fig. 4.8 Inverting amplifier

Substitute I_2 and I_1 into Eq. (4.9),

$$\frac{V_O - V_N}{R_2} + \frac{V_{IN} - V_N}{R_1} = 0. \tag{4.10}$$

Due to virtual ground rule,
$V_N = V_P = 0$ as shown in the circuit.
Replace V_N with zero in Eq. (4.9) and solve for the gain, V_O/V_{IN}.

$$\text{Gain} = \frac{V_O}{V_{IN}} = -\frac{R_2}{R_1}. \tag{4.11}$$

4.1.3.2 Biasing Op Amps

Op amp can either be powered by a dual rail supply ($\pm VDD$) or a single rail supply ($+VDD$). For the dual rail supply, the signal is centered around zero volts and swings between the positive and negative rails as shown in Fig. 4.9. In this case, the op amp needs to be biased at zero volts as this allows the maximum symmetrical swing.

The rule is to always bias the positive terminal of the op amp as shown in Figs. 4.10 and 4.12 and the bias voltage must be set at the level where the output has the maximum swing as shown in Fig. 4.11.

For the op amp with a single rail power supply, the bias voltage must be half of the power supply to guarantee maximum symmetrical swing as demonstrated in Fig. 4.11. The circuit shown in Fig. 4.12 has a voltage divider formed by two equal resistors, R_3 and R_4, to generate a bias voltage at half of the power supply rail, $+VDD$. For an ideal op amp, the positive and negative input voltages are equal. But this is not the case in the real world where there is always some small offset voltage between the two inputs. This offset voltage is in the range of microvolts and can be an issue for small signal detection and processing applications. In general, this offset

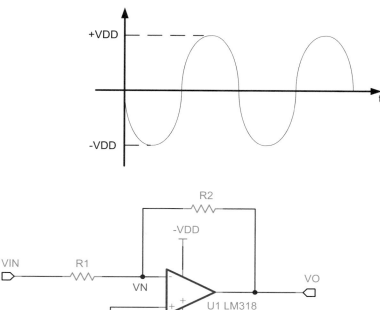

Fig. 4.9 Dual rail signal swing

Fig. 4.10 Op amp with dual supply rails

voltage is not a concern for video, audio, and communication designs, but it is good to keep it as low as possible. To minimize the offset voltage, set the parallel combination of R_3 and R_4 equal to R_2.

$$R_2 = R_3//R_4 = \frac{R_3 R_4}{R_3 + R_4},$$

and

$$R_3 = R_4 \quad \text{so } R_2 = \frac{R_3^2}{2R_3}.$$

$$\text{Therefore,} \quad R_3 = R_4 = 2R_2. \tag{4.12}$$

This bias voltage can also be generated by a voltage regulator. The advantage of the regulator is that it rejects the power supply noise. But the disadvantages are adding more cost and making the design and layout more complicated.

Fig. 4.11 Single rail signal swing

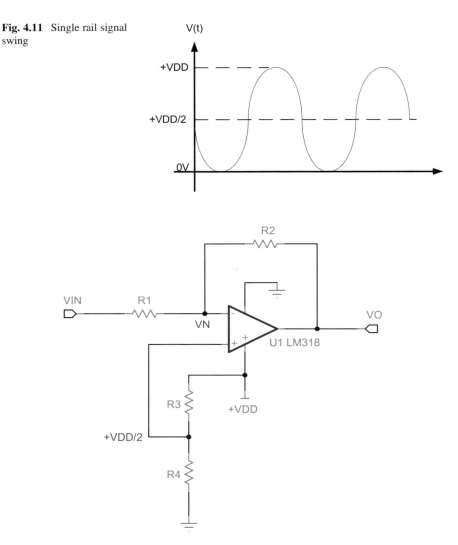

Fig. 4.12 Op amp with single rail supply

Again, only the positive terminal of the op amp needs to be biased at half of the power supply. Due to virtual ground rule, the negative DC voltage is at $+VDD/2$ and this also sets the output at $+VDD/2$. Now the whole circuit is DC balanced, which enables the signal to swing symmetrically around $+VDD/2$.

Another important rule to remember is that if a point in the circuit is connected to a DC voltage, the connection point becomes an AC or signal ground. So for a resistor divider circuit in Fig. 4.12, it is good to add a bypass capacitor C_1 in parallel with R_4 to provide a good AC ground as in Fig. 4.13. This capacitor does not affect the signal

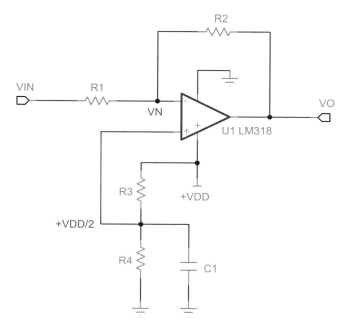

Fig. 4.13 Op amp with single rail supply

path at all, since the capacitor is on the positive input of the op amp which only has a DC bias voltage.

For AC signals, C_1 bypasses R_4 and provides a very low impedance path to ground. In this case, an RC filter is formed by R_3 and C_1 and the corner frequency is

$$f_{-3dB} = \frac{1}{2\pi R_3 C_1}. \tag{4.13}$$

In Eq. (4.13), it is preferable to select C_1 such that the -3 dB corner frequency is low enough to filter out the power supply noise. Now, let us bias a non-inverting amplifier circuit. The rule is the same as in the inverting case where only the positive input of the op amp needs to be biased. Figure 4.14 shows the biasing circuit where R_3 is connected to ground which is in the middle +VDD and −VDD rails. Again, to minimize the offset voltage, set R_3 equal to the parallel combination of R_1 and R_2. So,

$$R_3 = R_1//R_2 = \frac{R_1 R_2}{R_1 + R_2}, \tag{4.14}$$

where R_1 and R_2 determine the AC gain of the circuit as shown in Eq. (4.8).

Similarly, Fig. 4.15 shows a biasing circuit for the single rail supply non-inverting amplifier. The resistors R_3 and R_4 bias the positive input at half of the supply voltage. To minimize the offset voltage, set parallel combination of R_3 and R_4 equal to R_2, since DC current does not flow through R_1 because of the DC blocking capacitor C_3.

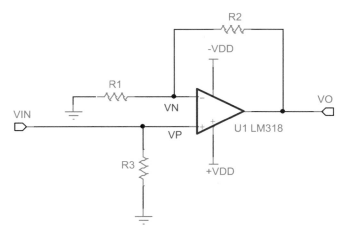

Fig. 4.14 Biasing op amp with dual rail supply

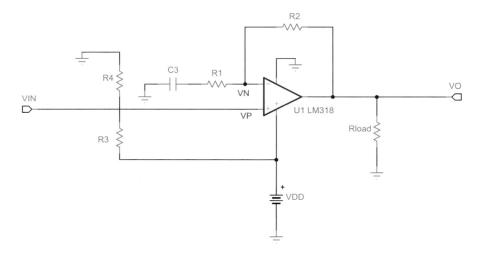

Fig. 4.15 Biasing op amp with single rail supply

$$R_3 // R_4 = R_2,$$

$$\frac{R_3 R_4}{R_3 + R_4} = R_2. \quad \text{Since } R_3 = R_4,$$

$$R_3 = R_4 = 2R_2, \tag{4.15}$$

where R_1 and R_2 determine the gain of the non-inverting amplifier circuit.

4.1.3.3 DC and AC-Coupled Op Amp Circuits

The next step is to isolate the DC bias voltages of the op amp to ensure that the input and output loadings do not change the DC voltages of the circuit. This is done by adding a DC blocking capacitor in series with the input and the output as shown in Fig. 4.16. The DC blocking or AC-coupling capacitor affects the frequency response of the circuit, so designers must perform the analysis to select the capacitor value such that the corner frequency is outside of the band of interest.

From the input V_{IN} looking into the circuit, the C_1 capacitor and resistors R_3 and R_4 in Fig. 4.17 form a high-pass filter and the corner frequency is

$$f_{-3dB} = \frac{1}{2\pi(R_2//R_3)C_1},$$

$$f_{-3dB} = \frac{R_2 + R_3}{2\pi(R_2R_3)C_1}. \tag{4.16}$$

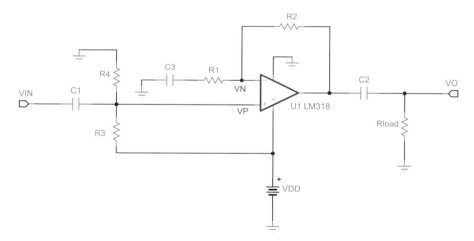

Fig. 4.16 Non-inverting AC-coupled input and output

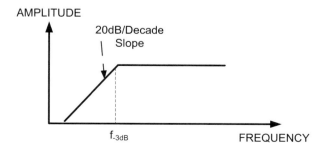

Fig. 4.17 Non-inverting high-pass filter response

The frequency response of the high-pass filter is shown in Fig. 4.17.The role of this capacitor C_3 is to block DC current from flowing through the resistor R_1 to keep all the DC voltages around the op amp equal. For AC signal, C_3 and R_1 also form a high-pass filter and its -3 dB corner is

$$f_{-3\text{dB}} = \frac{1}{2\pi R_1 C_3}. \tag{4.17}$$

For the output AC-coupling capacitor C_2, looking out from the op amp output, the capacitor C_2 and the resistor R_{load} form a high-pass filter and its -3 dB corner frequency is

$$f_{-3\text{dB}} = \frac{1}{2\pi R_{\text{load}} C_2}. \tag{4.18}$$

Again, C_2 must be selected to ensure that the corner frequency is lower than the lowest frequency of the band of interest.

Design Example 4.3 Design a non-inverting audio amplifier having the following specifications:

- Gain $= 2$
- Input -3 dB corner frequency $= 20$ Hz
- Output -3 dB corner frequency $= 20$ Hz
- From the input to output, there is a -6 dB attenuation at 20 Hz. This is due to cascading -3 dB input and -3 dB output stages
- Input impedance is 10K Ω
- Output load impedance is 20K Ω
- AC-coupled input and output
- $+12$ V single rail power supply

Simulate [2] the circuit to verify the results.
Solution:
Use the circuit diagram in Fig. 4.16 and calculate all the component values.

$$\text{Gain} = \frac{V_O}{V_{\text{IN}}} = \frac{R_2}{R_2} + 1 = 2,$$

So, $R_2 = R_1$.
Let $R_2 = R_1 = 20$K Ω, reasonable value for audio design.
From Eq. (4.15),

$$R_3 = R_4 = 2R_2 = 2(20\text{K}) = 40\text{K } \Omega.$$

The input impedance is R_3 in parallel with R_4 which is

$$\frac{(40K)(40K)}{40K + 40K} = 20K \ \Omega.$$

From Eq. (4.13), the -3 dB corner is

$$20 \ \text{Hz} = \frac{R_2 + R_3}{2\pi(R_2 R_3)C_1},$$

$$C_1 = \frac{40K + 40K}{2\pi(20)(40K)(40K)} = 0.4 \ \mu\text{F}.$$

Let C_1 be 0.47 μF for 0.47 μF being a standard capacitor value. Calculate the output AC-coupling capacitor C_2 using Eq. (4.18).

$$20 \ \text{Hz} = \frac{1}{2\pi R_{\text{load}} C_2},$$

where R_{load} is 20K Ω.

$$C_2 = \frac{1}{2\pi(20)(20K)} = 0.4 \ \mu\text{F}.$$

Let C_2 be 0.47 μF for 0.47 μF being a standard capacitor value. Calculate the DC blocking capacitor C_3 using Eq. (4.17).

$$20 \ \text{Hz} = \frac{1}{2\pi R_1 C_3},$$

$$C_3 = \frac{1}{2\pi(20)(20K)} = 0.4 \ \mu\text{F}.$$

Let C_3 be 0.47 μF for 0.47 μF being a standard capacitor value.

The final circuit is shown in Fig. 4.18 and simulation in Fig. 4.19.

The simulation in Fig. 4.19 verified that there is a -6 dB attenuation at 20 Hz; the overall circuit has a gain of 2 or $+6$ dB and, at 20 Hz, the gain is $+6$ dB -6 dB attenuation. Therefore, the graph in Fig. 4.19 shows a 0 dB gain at 20 Hz.

Now for the inverting circuit shown in Fig. 4.20, from the input V_{IN} looking into the circuit, the gain of the circuit is modified by the impedance of the capacitor and is equal to R_2 divided by R_1 plus the impedance of the capacitor C_2.

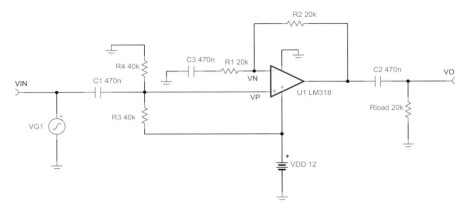

Fig. 4.18 Final circuit for Example 4.1

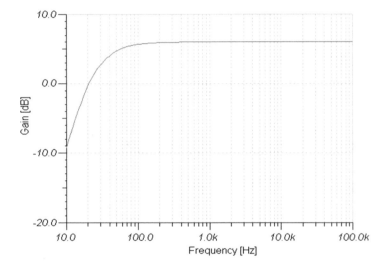

Fig. 4.19 Final circuit simulation for Example 4.1

$$\frac{V_O}{V_{IN}} = -\frac{R_2}{R_1 + Z_2},\tag{4.19}$$

where Z_2 is the impedance of C_2 and is equal to $\frac{1}{2\pi f C_2}$ and f is frequency.

Substitute Z_2 into Eq. (4.19). The magnitude of the gain, ignoring the negative sign as the sign only indicates the output is inverted, is

$$\frac{V_O}{V_{IN}} = \frac{2\pi f C_2 R_2}{2\pi f C_2 R_1 + 1}.\tag{4.20}$$

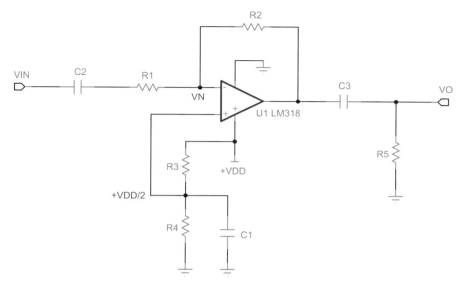

Fig. 4.20 Inverting AC-coupled input and output

For the corner frequency of

$$f_{corner} = \frac{1}{2\pi R_1 C_2}, \tag{4.21}$$

where C_2 and R_1 are the input impedance looking into the circuit. Substitute the corner frequency into Eq. (4.20) and the magnitude of the gain at this frequency becomes

$$\frac{V_O}{V_{IN}} = \frac{R_2}{2R_1}. \tag{4.22}$$

For R_2 equal to R_1, the gain at the corner frequency is

$$20 \log (0.5) = -6 \text{ dB}.$$

So, the AC-coupling capacitor at the input of the inverting amplifier forms a high-pass filter where the corner frequency shown in Eq. (4.21) is at -6 dB as demonstrated in Fig. 4.21. As frequency gets higher and higher, the gain in Eq. (4.20) is dominated by the resistors R_2 and R_1, which lead to the same response as the inverting amplifier without the AC-coupling capacitor. Again, designers need to select C_2 so that the corner frequency is lower than the lowest frequency of the signal.

C_3 in Fig. 4.20 is calculated the same way as in the non-inverting case.

Fig. 4.21 Inverting high-pass filter response

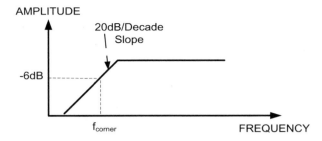

Design Example 4.4 Design an inverting audio amplifier having the following specifications:

- Gain $= -2$
- Input -3 dB corner frequency $= 20$ Hz
- Output -3 dB corner frequency $= 20$ Hz
- From the input to output, there is a -9 dB attenuation at 20 Hz. This is due to cascading -6 dB input and -3 dB output stages
- Input impedance is greater than 10K Ω
- Output load impedance is 20K Ω
- AC-coupled input and output
- +12 V single rail power supply

Simulate [2] the circuit to verify the results.
Solution:
Use the circuit diagram in Fig. 4.20 and calculate all the component values.

$$\text{Gain} = \frac{V_O}{V_{\text{IN}}} = \frac{R_2}{R_1} = 2 \text{ (neglect minus sign as it only indicates the phase)},$$

So, $R_2 = 2R_1$.
Let $R_1 = 20$K so $R_2 = 2(20\text{K}) = 40$K.
From Eq. (4.12),

$$R_3 = R_4 = 2R_2 = 2(40\text{K}) = 80\text{K}.$$

The input impedance is equal to R_1 which is 20K since the positive and negative inputs of the op amp are equal, AC ground.
From Eq. (4.21), the input corner frequency (-6 dB) is

$$20 \text{ Hz} = \frac{1}{2\pi R_1 C_2},$$

$$C_2 = \frac{1}{2\pi(20)(40\text{K})} = 0.2 \text{ μF}.$$

Let C_2 be 0.22 μF for 0.22 μF being a standard capacitor value.

Calculate the output AC-coupling capacitor C_3 using Eq. (4.18).

$$20 \text{ Hz} = \frac{1}{2\pi R_5 C_3},$$

where R_5 (load) is 20K Ω.

$$C_3 = \frac{1}{2\pi(20)(20\text{K})} = 0.4 \text{ µF}.$$

Let C_3 be 0.47 µF for 0.47 µF being a standard capacitor value.
Calculate the bypass capacitor C_1 using Eq. (4.13).

$$f_{-3\text{dB}} = \frac{1}{2\pi R_3 C_1} = 20 \text{ Hz},$$

this is the filter corner for the bias voltage.

$$C_1 = \frac{1}{2\pi(20)(80\text{K})} = 0.1 \text{ µF}.$$

The final circuit is shown in Fig. 4.22 and simulation in Fig. 4.23.
The simulation in Fig. 4.23 verified that there is a -9 dB attenuation at 20 Hz; the overall circuit has a gain of 2 or $+6$ dB and, at 20 Hz, the gain is $+6$ dB -9 dB attenuation. Therefore, the graph in Fig. 4.23 shows roughly a -3 dB gain at 20 Hz.

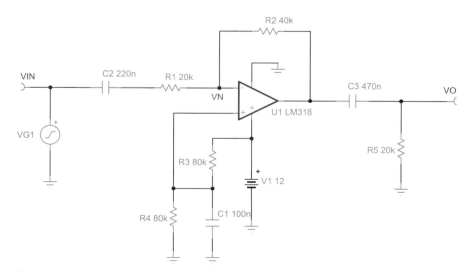

Fig. 4.22 Final circuit for Example 4.2

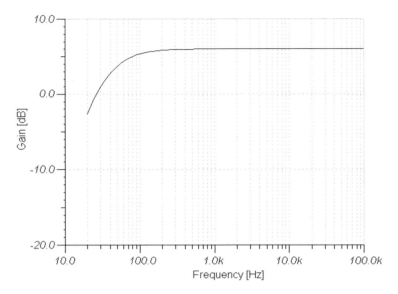

Fig. 4.23 Final circuit simulation for Example 4.2

4.1.3.4 First Order Active Filter Design

Let us assume that the active filters are running on a single rail power supply. This is more common in todays electronics as it is more expensive to implement a design powered by a dual rail power supply.

There are two topologies for the first-order lowpass filter design, inverting lowpass and non-inverting lowpass. For the non-inverting topology, the gain must be greater than 1 as shown in the previous section, and, for the inverting topology, the output is 180° phase shifted from the input.

To create an inverting lowpass filter, simply take the circuit in Fig. 4.22 and add a capacitor in parallel with the resistor R_2. The new circuit shown in Fig. 4.24 has the upper -3 dB corner of

$$f_{-3\text{dB}} = \frac{1}{2\pi R_2 C_4}.\tag{4.23}$$

All other design parameters and methodologies are the same as those shown in the previous op amp design sections.

Design Example 4.5 Design a first-order inverting audio filter having the same specifications as in Design Example 4.4 but adding an upper frequency limitation at 20 kHz. Simulate the circuit to verify the results.

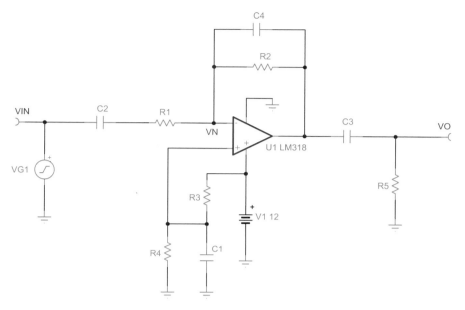

Fig. 4.24   First-order active lowpass filter

Solution:
Use the circuit in Fig. 4.24. From Eq. (4.23), the upper -3 dB frequency is

$$20 \text{ kHz} = \frac{1}{2\pi R_2 C_4}.$$

From Design Example 4.2, R_2 is 40K Ω. So,

$$C_4 = \frac{1}{2\pi(40\text{K})(20 \text{ kHz})} = 200 \text{ pF}.$$

The final circuit and simulation are shown in Figs. 4.25 and 4.26, respectively. It is verified that the circuit frequency response has an upper frequency limitation at 20 kHz. This is the first-order filter circuit with the -3 dB frequency at 20 kHz and the slope decaying at 20 dB per decade.

As mentioned earlier, the output of the inverting lowpass filter has a 180° phase shift as compared to the input. Figure 4.27 shows the simulated gain of 2 and phase relationship of the input and output waveforms.

Now to realize the non-inverting first-order lowpass filter circuit, take the non-inverting amplifier circuit and add a capacitor in parallel to R_2 to limit the op amp bandwidth and an RC filter at the positive input of the op amp as shown in Fig. 4.28. Op amp typically has a high gain bandwidth product and can go unstable if the bandwidth is not limited to the operating frequency range. Doing stability

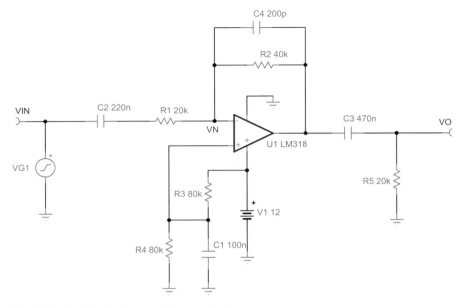

Fig. 4.25 Final final-order active lowpass circuit

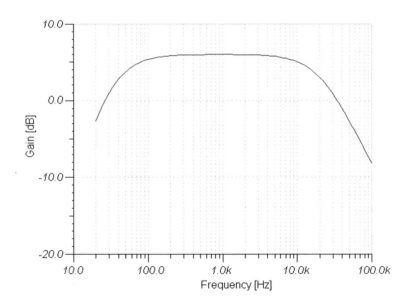

Fig. 4.26 Final first-order active filter simulation

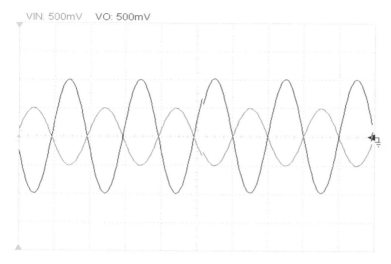

Fig. 4.27 Input versus output waveforms

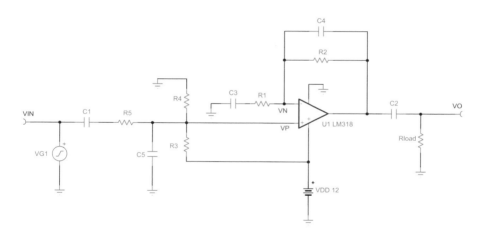

Fig. 4.28 Non-inverting first-order lowpass

analysis is beyond the scope of this book but designers can learn more in [3]. This new circuit is shown in Fig. 4.28 and the -3 dB corner is dominated by the resistor R_5 and the capacitor C_5,

$$f_{-3\text{dB}} = \frac{1}{2\pi R_5 C_5}.$$
(4.24)

With C_4 connected in parallel with resistor R_2, the op amp bandwidth is limited to the -3 dB corner set by this RC filter. The corner frequency is

$$f_{-3dB} = \frac{1}{2\pi R_2 C_4}. \tag{4.25}$$

Also, in Fig. 4.28, the filter resistor R_5 needs to be one-tenth of the parallel combination of the resistors R_3 and R_4. This is to minimize the effects of the voltage divider formed by R_5 and the parallel combination of R_3 and R_4.

Design Example 4.6 Design a first-order non-inverting audio filter having the same specifications as in Design Example 4.3 but adding an upper frequency limitation at 20 kHz. Simulate the circuit to verify the results.
Solution:
Use the circuit in Fig. 4.28. From Eq. (4.24), the upper -3 dB frequency is

$$20 \text{ kHz} = \frac{1}{2\pi R_2 C_5}.$$

From Design Example 4.3, $R_3 = R_4 = 40\text{K } \Omega$. So,

$$R_5 = \frac{1}{10}(R_3//R_4) = \frac{1}{10}\frac{R_3 R_4}{(R_3 + R_4)} = 2\text{K},$$

$$C_5 = \frac{1}{2\pi(20\text{K})(2 \text{ kHz})} = 0.00398 \text{ } \mu\text{F}.$$

Now, since the maximum upper frequency is 20 kHz, let us limit the op amp to 40 kHz which is two times the signal bandwidth. This provides plenty of bandwidth margins.
From Eq. (4.25),

$$40 \text{ kHz} = \frac{1}{2\pi R_2 C_4},$$

In Design Example 4.3, $R_2 = 20\text{K}$ so

$$C_4 = \frac{1}{2\pi(20\text{K})(40 \text{ kHz})} = 200 \text{ pF}.$$

The final circuit and simulation are shown in Figs. 4.29 and 4.30, respectively. It is verified that the circuit frequency response has an upper frequency limitation at 20 kHz. This is the first-order non-inverting filter circuit with the -3 dB frequency of 20 kHz and a slope of -20 dB/decade.

Figure 4.31 shows the simulated gain of 2 and phase relationship of the input and output waveforms. The simulation results are correct and correlated with the calculations very well.

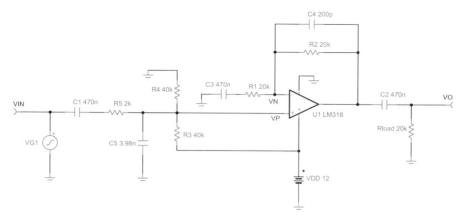

Fig. 4.29   Final non-inverting first-order lowpass

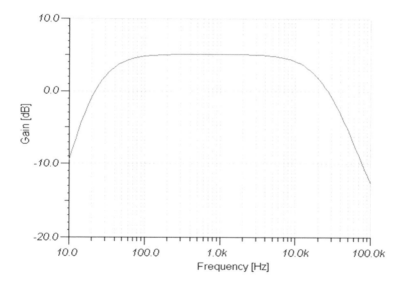

Fig. 4.30   Final non-inverting first-order response

4.1.3.5   Operational Amplifier (Op Amp) Fundamentals

The best and the most popular topology for a second-order lowpass design is Sallen-Key circuit [4]. The three different Sallen-Key circuits are unity gain, gain of 2, and gain of higher than 2. The higher the gain the more unstable the circuit becomes, so it is best to keep the overall gain of 2 or less. If more gain is required for the design, then use another op amp as a gain stage to boost the overall gain. This guarantees good stability.

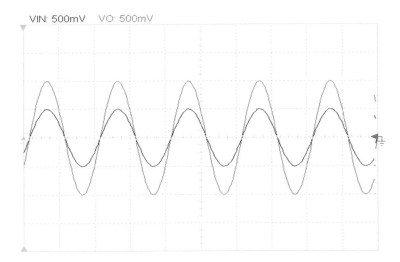

Fig. 4.31 Non-inverting input versus output waveforms

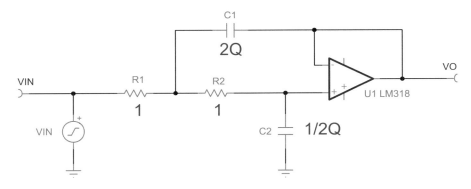

Fig. 4.32 Sallen-Key circuit with Gain = 1

Sallen-Key Circuit with Gain = 1

Figure 4.32 shows a second-order lowpass filter with a gain of 1 [4]. The resistor and capacitor values are normalized to 1 and Q where Q determines the gain peaking or GP at the corner frequency as shown in Fig. 4.33. The gain peaking or GP equation is

$$\text{GP} = 20 \log_{10} \frac{2Q^2}{\sqrt{4Q^2 - 1}}, \quad \text{for } Q > 0.707. \tag{4.26}$$

There is no gain peaking for $Q = 0.707$.

To calculate the values of the capacitors and resistors, use the following equations.

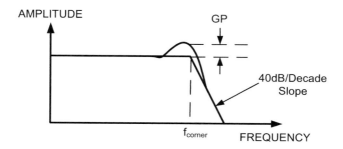

Fig. 4.33 Sallen-Key response and GP

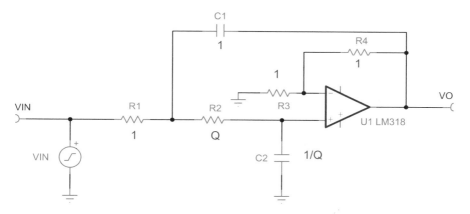

Fig. 4.34 Sallen-Key circuit with Gain = 2

$$R_{new} = K_m R_{old,} \tag{4.27}$$

where K_m is the new resistance and R_{old} is the normalized resistance shown in Fig. 4.32.

$$C_{new} = \frac{1}{K_f K_m} C_{old}, \tag{4.28}$$

where K_f is the corner frequency in rad/s, $K_f = 2\pi f; f$ is the corner frequency in Hertz.

Sallen-Key Circuit with Gain = 2

Since R_3 and R_4 are equal in Fig. 4.34 [4], the overall circuit gain is 1 plus the ratio of resistor R_4 over resistor R_3 or 2 as shown in the non-inverting amplifier section. All the values in the circuit are calculated by the same methods in the Sallen-Key Circuit with Gain = 1 section.

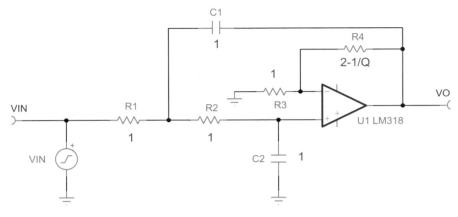

Fig. 4.35 Sallen-Key circuit with Gain = 3 − 1/Q

Sallen-Key Circuit with Gain = 3 − 1/Q

The overall gain in Fig. 4.35 [4] is

$$1 + \frac{R_4}{R_3} = 1 + \frac{2 - 1/Q}{1} = 3 - \frac{1}{Q}. \qquad (4.29)$$

All the values in the circuit are calculated by the same methods demonstrated in the Sallen-Key Circuit with Gain = 1 section.

Design Example 4.7 Design a non-inverting second-order Sallen-Key lowpass filter having the following specifications:

- Gain = 2
- Corner frequency = 20 kHz
- Gain peaking = 5 dB
- Output load impedance = 20K Ω
- +12 V single rail power supply
- Use the same input and output coupling capacitors as in Design Example 4.3.

Simulate [2] the circuit to verify the results.
Solution:
For Gain = 2, use the circuit in Fig. 4.34.
For corner frequency = 20 kHz,

$$K_f = 2\pi f = 2\pi(20 \text{ kHz}) = 125,663.7 \text{ rad/s}.$$

Let $K_m = 2 \times 10^4$.
For gain peaking = 4 dB,

$$4\,\mathrm{dB} = 20 \log_{10} \frac{2Q^2}{\sqrt{4Q^2 - 1}},$$

Now, solve for Q and Q is 1.5.

$$R_{\mathrm{new}} = K_{\mathrm{m}} R_{\mathrm{old}} = 2 \times 10^4 = R_1 = R_3 = R_4.$$

$$R_{2\,\mathrm{old}} = Q = 1.5,$$

$$R_2 = K_{\mathrm{m}} R_{\mathrm{old}} = 2 \times 10^4\,(1.5) = 30\mathrm{K}.$$

$$C_{1\,\mathrm{old}} = 1,$$

$$C_{1\,\mathrm{new}} = \frac{1}{K_{\mathrm{f}} K_{\mathrm{m}}} C_{\mathrm{old}} = \frac{1}{(125,663.7\,)(2 \times 10^4)} = 398\ \mathrm{pF}.$$

$$C_{2\,\mathrm{old}} = 1/Q = 0.667,$$

$$C_{2\,\mathrm{new}} = \frac{1}{K_{\mathrm{f}} K_{\mathrm{m}}} C_{\mathrm{old}} = \frac{0.667}{(125,663.7\,)(2 \times 10^4)} = 265\ \mathrm{pF}.$$

The final circuit is shown in Fig. 4.36 where C_3, C_5, and C_4 came from the Design Example 4.3. R_6 and R_7 form a voltage divider to bias the op amp at half of the power supply voltage. The parallel combination of the resistors R_6 and R_7 is selected to be 10 times larger than the total resistance of R_1 and R_2. This guarantees that the bias resistors will not affect the overall gain.

The simulations in Figs. 4.36 and 4.37 show the following results:

- Gain peaking $= 4$ dB
- Corner frequency $= 20$ kHz

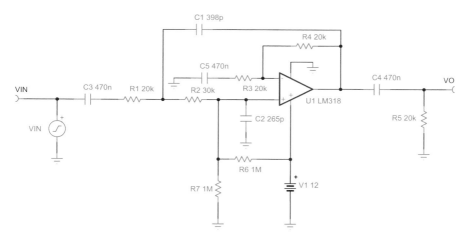

Fig. 4.36 Final Sallen-Key circuit with Gain $= 2$

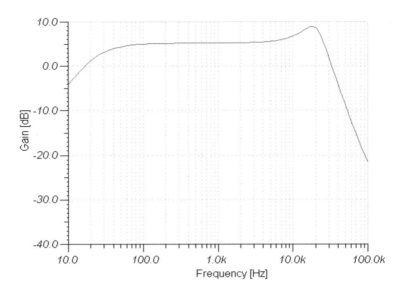

Fig. 4.37 Final Sallen-Key circuit simulation

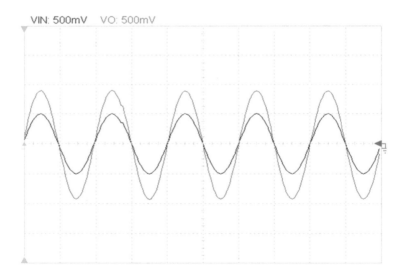

Fig. 4.38 Final Sallen-Key input versus output

- Slope $= -40$ dB/decade, second-order lowpass filter
- Passband signal gain $= 1.85$ instead of 2. This is due to the bias resistors R_6 and R_7 loading the signal down. Increasing the total R_6 and R_7 resistance reduces the effect but will cause problems with not having adequate bias current required for the op amp. Another option is reducing the total resistance of R_1 and R_2. This is a better option, but it requires a total redesign of the filter (Fig. 4.38).

4.2 Summary

As demonstrated throughout this chapter, filter designs are complicated and require doing thorough system analysis using a circuit simulator such as [2]. A filter topology is selected based on the following criteria:

- Is the gain greater than 1? If yes, then an active filter is needed. If no, then either an active or a passive filter can be selected.
- How much attenuation is needed to prevent aliasing for the ADC input and to reject the sampling noise for the DAC output? This determines first, second, or higher order filter topology.
- Is maintaining the input and output phase important? If yes, then only a non-inverting filter can be used.
- For active filters, is the circuit running on a single or dual rail power supply? Refer to the DC and AC Couple and Biasing Op Amp sections, and bias the circuit to allow for a maximum symmetrical swing.
- For passive filters, is inductor shielding required? Inductors tend to radiate high-frequency noise, so shielding may be required to contain the radiation. In summary, DSP systems require analog filters to limit the signal bandwidth before sampling, to reconstruct the analog signal from the DAC output, and to eliminate analog noise modulated on the signal. These filters are critical, and proper design techniques outlined in this chapter should be followed to achieve the performance goals.

References

1. Texas Instruments Inc., Understanding Data Converters, Mixed-Signal Products, SLAA013 (1995)
2. Texas Instruments Inc., Spice-Based Analog Simulation Program (2008). http://focus.ti.com/docs/toolsw/folders/print/tina-ti.html
3. F. Sergio, *Design with Operational Amplifiers and Analog Integrated Circuits* (McGraw-Hill, New York, 2002)
4. M.E. Van Valkenburg, *Analog Filter Design* (Holt, Rinehart and Winston, New York, 1982)

Chapter 5
Data Converter Overview

This chapter provides an overview of analog-to-digital and digital-to-analog converters and their applications in audio and video systems design. There are many factors affecting the performance of the converters and these can be minimized if designers understand the converter's sampling techniques and quantization noise, the necessity of having input and output filters, and the proper system design and layout.

5.1 CPU/DSP System

Figure 5.1 shows a typical DSP system where the input and output are analog and data converters plus processing elements reside in the middle of the signal chain. The theories and applications of the input Gain Stage, Anti-Aliasing Filter (ADC input), and Reconstruction Filter (DAC output) are covered in Chap. 4. This chapter focuses on the ADC and DAC blocks of the DSP Signal Chain in Fig. 5.1.

In general, a DSP system captures an analog input, amplifies the signal, band-limits the signal for sampling, converts it to digital, processes the data in digital domain, converts it back to analog, and filters the sampling noise to reconstruct the analog signal.

The design goal is to maintain or improve the signal quality as it propagates through all the blocks shown in Fig. 5.1. The question is why is it necessary to convert the analog signal to digital and process it in digital domain? It is because:

- There are no linear and nonlinear distortions.
- Data compression is possible in digital domain. This is key for video, audio, and communication as these systems have limited transmission bandwidth.
- It is easy to upgrade the system by replacing the software and or DSP algorithms.

© The Author(s), under exclusive license to Springer Nature Switzerland AG 2023
T. T. Tran, *High-Speed System and Analog Input/Output Design*,
https://doi.org/10.1007/978-3-031-04954-5_5

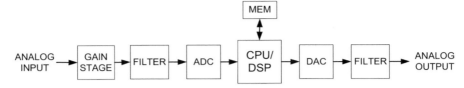

Fig. 5.1 CPU/DSP signal chain

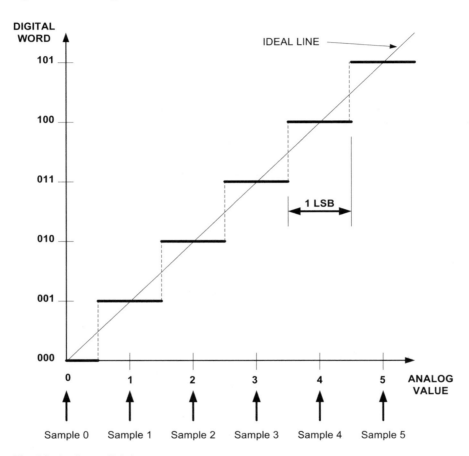

Fig. 5.2 Analog-to-digital converter

5.2 Analog-to-Digital Converter (ADC)

An ADC converts the analog input to the digital output word by sampling and comparing the analog level to the digital word. This sampling point is to decide what digital code is equivalent to this analog value. For example, in Fig. 5.2, sampling

Table 5.1 ADC input and output

Digital code	Analog range (V)	Quantization error
000	0–0.5	1/2 LSB or 0.5 V
001	0.5–1.5	1 LSB or 1 V
010	1.5–2.5	1 LSB or 1 V
011	2.5–3.5	1 LSB or 1 V
100	3.5–4.5	1 LSB or 1 V
101	4.5–5.5	1 LSB or 1 V
110	5.5–6.5	1 LSB or 1 V
111	6.5–7.5	1 LSB or 1 V

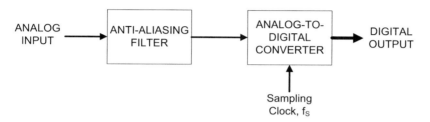

Fig. 5.3 Analog-to-digital block diagram

point 0 occurs at 1 V analog level translating to 001 digital output word, sampling point 1 occurs at 2 V analog level translating to 010 digital output word, and so on.

In Fig. 5.2, 1 LSB is defined as one Least Significant Bit and at every sampling point; the analog level can vary from $\pm 1/2$ LSB from the center. This is known as a quantization error. The LSB voltage, V_{LSB}, is equal to

$$V_{\text{LSB}} = \frac{V_{\text{ref}}}{2^N}, \tag{5.1}$$

where V_{ref} is the reference voltage and N is the number of bits.

For the 3-bit ADC and V_{ref} of 8 V, the V_{LSB} is equal to 1 V. Table 5.1 shows an example of a 3-bit ADC sampling an 8 V analog input signal.

Overall, the equation for calculating the voltages of the ADC [1] is as follows:

$$V_{\text{ref}}\left(b_1 2^{-1} + b_2 2^{-2} + b_3 2^{-3} + \cdots + b_N 2^{-N}\right) = V_{\text{in}} \pm V_{\text{x}}, \tag{5.2}$$

where V_{x} is $-\frac{1}{2}V_{\text{LSB}} \leq V_X \leq \frac{1}{2}V_{\text{LSB}}$ and b_1 is the most significant bit and b_N is the least significant bit.

Figure 5.3 shows a practical ADC block diagram where the analog input must be band-limited before being converted to digital word. This is because the Nyquist sampling theory defined that the sampling clock must be at least two times the analog bandwidth to prevent aliasing. Aliasing is the image of the analog signal folded back into the frequency of interest; aliasing degrades the ADC performance. Therefore, an anti-aliasing filter must be placed at the input of the ADC.

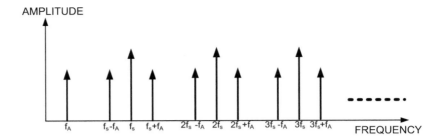

Fig. 5.4 Frequency spectrum of a sampled signal

Fig. 5.5 Aliasing

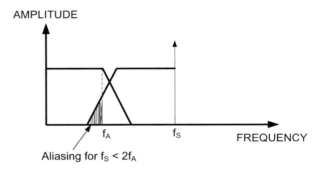

5.2.1 Sampling

An ADC utilizes the sampling clock, f_s, to sample the analog input and represents the level in a digital word as shown in the previous section. Sampling is equivalent to amplitude modulating the signal, f_A, into a carrier equal to the sampling frequency and generates a frequency spectrum shown in Fig. 5.4.

Nyquist Theorem states that the sampling frequency f_s must be equal to or greater than two times the signal bandwidth f_A or

$$f_s \geq 2f_A. \tag{5.3}$$

If the sampling frequency is less than two times the signal bandwidth, then the sampled signal aliases back into the signal bandwidth causing the dynamic range degradation. Figure 5.5 shows an aliasing region when this is the case. So, to guarantee compliant to Nyquist Theorem, the ADC must have an anti-aliasing filter at the input to limit the signal bandwidth and a sampling frequency greater than two times the signal bandwidth. One important rule-of-thumb for selecting an anti-aliasing filter is that higher sampling frequency ADC requires a lower order anti-aliasing filter. This is because higher sampling frequency pushes the noise spectrum further away from the analog spectrum and enables the lower order filter to prevent

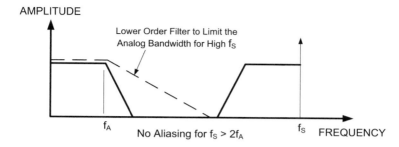

Fig. 5.6 Higher sampling frequency

the noise image from aliasing back into the band of interest. Figures 5.5 and 5.6 demonstrate this phenomenon.

Todays technologies enable very high sampling frequency ADC, so it is not necessary to have higher than a second-order anti-aliasing filter at the input of the ADC.

5.2.2 Quantization Noise

Quantization process is a process of sampling an analog signal and representing the signal in a sequence of digital bits. The issue is that the input signal varies rapidly and generates a quantization error where a particular bit oscillates back and forth between two digital levels. For example, in Fig. 5.7 at Sample 2 and Sample 3 points, the digital word for these two levels could be 000 or 001, since these points are halfway between two levels and at a given sample time, the quantizer can interpret this level as a 000 or 001. Therefore, the quantization error is $\pm q/2$ where q is a quantization step equal to 1 LSB.

Now assume that the quantization error voltage between the quantized levels and the sampled voltage is uniformly distributed between $-q/2$ and $+q/2$ where q is equal to 1 LSB voltage. In this case, the Probability Density Function, $f_Q(x)$, is shown in Fig. 5.8 where the area under the curve is equal to one.

The quantization noise in RMS value is

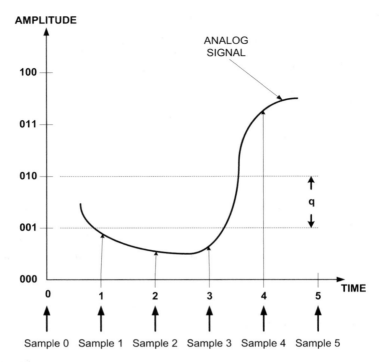

Fig. 5.7 Quantization error

Fig. 5.8 Probability density
function, $f_Q(x)$

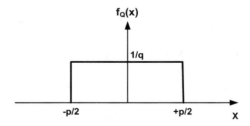

Fig. 5.9 Sinusoidal input waveform

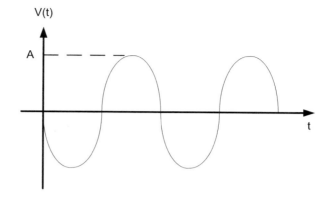

$$V_Q(\text{rms}) = \left[\int_{-q/2}^{q/2} x^2 f_Q(x) dx \right]^{1/2}$$

$$= \left[\frac{1}{q} \int_{-q/2}^{q/2} x^2 dx \right]^{1/2} = \left[\frac{1}{q} \int_{-q/2}^{q/2} x^2 dx \right]^{1/2} \qquad (5.4)$$

$$= \left[\frac{1}{q} \int_{-q/2}^{q/2} x^2 dx \right]^{1/2} = \left[\frac{1}{q} \left[\frac{1}{3} x^3 \right]_{-q/2}^{+q/2} \right]^{1/2}$$

$$= \frac{q}{\sqrt{12}}.$$

The RMS value of a sinusoidal signal is

$$V_{IN}(\text{rms}) = \left[\frac{1}{T} \int_{0}^{T} V^2(t) dt \right]^{1/2},$$

where $V(t) = A\cos(2\pi f_c t)$ and A is the zero-to-peak voltage as shown in Fig. 5.9. Therefore,

$$V_{IN}(\text{rms}) = \left[\frac{1}{T} \int_{0}^{T} A^2 \cos^2(2\pi f_c t) dt \right]^{\frac{1}{2}}.$$

Fig. 5.10 V_{REF} and peak-to-peak voltage

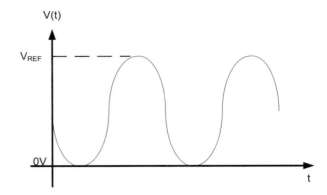

Since

$$\cos^2(2\pi f_c t) = \frac{1 + \cos(4\pi f_c t)}{2},$$

$$V_{IN}(\text{rms}) = \left[\frac{A^2}{T}\int_0^T \frac{1 + \cos(4\pi f_c t)}{2} dt\right]^{1/2} = \left[\frac{A^2}{T}\frac{T}{2}\right]^{1/2} = \frac{A}{\sqrt{2}}. \qquad (5.5)$$

A is equal $1/2V_{ref}$ since V_{ref} is equal to the peak-to-peak voltage of a sinusoidal wave as shown in Fig. 5.10 and A is the zero-to-peak value.

Now, substitute $A = 1/2V_{ref}$ into Eq. (5.5),

$$V_{IN(\text{rms})} = \frac{V_{ref}}{2\sqrt{2}}, \qquad (5.6)$$

or

$$V_{ref} = V_{peak\text{-to-peak}} = 2.828 V_{IN(\text{rms})}.$$

The signal-to-noise or SNR is defined as the log of the ratio of the input RMS voltage over the quantization noise.

$$\text{SNR} = 20\log_{10}\frac{V_{IN}(\text{rms})}{V_Q(\text{rms})}. \qquad (5.7)$$

Substitute Eqs. (5.6) and (5.4) into Eq. (5.7),

$$\text{SNR} = 20\log_{10}\left(\frac{V_{ref}}{2\sqrt{2}}\frac{\sqrt{12}}{q}\right).$$

Since $q = V_{LSB}$ and from Eq. (5.1)

$$V_{LSB} = \frac{V_{ref}}{2^N},$$

$$\text{SNR} = 20 \log_{10}\left(\frac{\sqrt{3}}{\sqrt{2}} 2^N\right)$$

$$= 6.02N + 1.76 \text{ dB}.$$

(5.8)

Equation (5.8) indicates that the performance of an ADC depends on the number of bits used to quantize the analog signal. It is roughly 6 dB/bit. This equation was derived assuming that the only error in the system is quantization error. In the real world, other factors such as power supply noise and clock jitter generate additional errors that degrade the signal-to-noise significantly. So, another equation to measure the overall performance of the ADC is

$$\text{ENOB, Effective Number of Bits} = \frac{\text{SNR} - 1.76}{6.02}.$$

(5.9)

For example, if a 16-bit ADC has an SNR specification of 86 dB, the Effective Number of Bits is

$$\text{ENOB} = \frac{86 - 1.76}{6.02} = 14.$$

What this means is that the 16-bit ADC only performs at a 14-bit level due to quantization noise and other system-related noise, such as power supplies, clocks, and others degrading its performance.

5.2.3   ADC Digital Output Waveforms

As silicon technology advances to enable faster switching and higher data rate, the digital waveforms also must change to be compatible with low voltage, low power devices. The switching logic levels for IO (inputs and outputs) used to be 3.3 V, and now these logic levels are as low as 1.2 V. Reducing the switching voltage significantly lowers the power consumption as defined in Eq. (5.1).

$$\text{Dynamic Power} = P_{dyn} = CV^2f,$$

(5.10)

where C is the capacitive load, V is the switching voltage, and f is the switching frequency. For digital logic, the input impedance of a gate is very high, so the timing is mostly affected by the input gate capacitance. From Eq. (5.1), the most effective way to reduce the power consumption of a system is to lower the switching voltage.

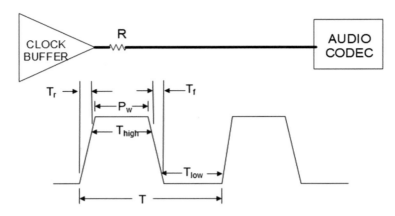

Fig. 5.11 Digital-to-analog

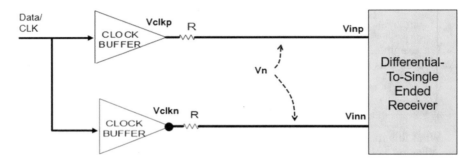

Fig. 5.12 Differential signaling scheme

For example, lowering the switching voltage by half reduces the dynamic power by four. Many battery-powered portable systems apply voltage and frequency scaling techniques to extend the battery life by lowering both the voltage and frequency when the system is not active.

Traditionally, Fig. 5.11, digital signal transmission scheme is a single ended scheme in which the timings are based on input high voltage (V_{ih}), input low voltage (V_{il}), rise time, fall time, and duty cycle (P_w/T). For high-speed systems, the trend is going to differential signal scheme as shown in Fig. 5.12 which has a positive going signal and a negative going signal transmitted simultaneously.

The advantages of differential signaling over single ended signaling are:

- High or low state does not require the signal to switch from V_{il} to V_{ih}. The signal only has to go above or below the common voltage as shown in Fig. 5.11. This significantly lowers the dynamic power as described.
- Higher common mode noise cancellation. The two signals within one differential pair must be routed the exact same way, next to each other, one time or two times

the width of the trace. Since both signals are the same, the noise coupled to one is similar to noise coupled to the other signal. This noise on both positive and negative signals is referred as common mode noise. This common mode noise is generally rejected by the receiver. The receiver is typically being implemented using a differential to single ended amplifier, and its output is

$$V_{out} = \text{Gain}(V_{inp} - V_{inn}),$$ (5.11)

where Gain is the amplifier's gain, V_{inp} is the positive input signal, and V_{inn} is the negative input signal.

Let V_n be noise voltage coupled to both positive and negative signals in one differential pair. In this case,

$$V_{inp} = V_{clkp} + V_n,$$ (5.12)

$$V_{inn} = V_{clkn} + V_n,$$ (5.13)

where V_{clkp} and V_{clkn} are clock signals without noise.
Now, substitute Eqs. (5.12) and (5.13) into Eq. (5.11).

$$V_{out} = \text{Gain}(V_{clkp} + V_n - V_{clkn} - V_n),$$

$$V_{out} = \text{Gain}(V_{clkp} - V_{clkn}).$$ (5.14)

Equation (5.14) shows that the differential receiver cancels common mode noise coupled to the transmitted signals. This is a major advantage for high-speed signaling.

Another important concept in high-speed design is understanding the eye diagram. Eye diagram is a composite signal which has important timing parameters embedded in the diagram. For example, Fig. 5.13 shows an eye diagram of the differential to single ended receiver output. Figure 5.14 eye diagram is being used to show data signal distortion, noise, and signal attenuation due to transmission line effects and intersymbol interference (ISI). ISI is the effect of adjacent pulses spilling over the neighboring pulses.

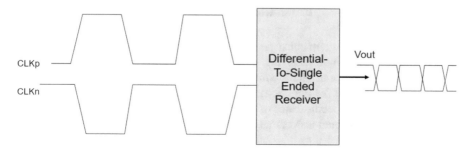

Fig. 5.13 Differential to single ended output eye

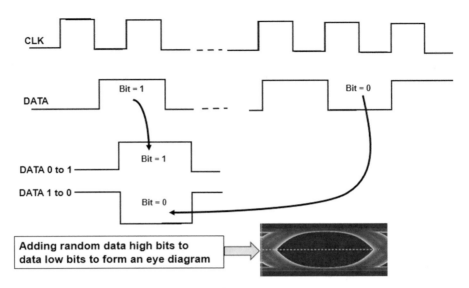

Fig. 5.14 Random data eye diagram [2]

Fig. 5.15 Eye diagram
timing information

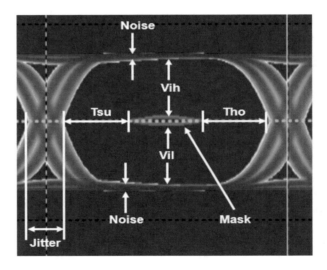

Figure 5.15 Eye Diagram consists of important timing parameters of a high-speed system in transmitting and receiving serial digital data. The industry uses a pseudo random binary sequence or PRBS which has a maximum length of $2^n - 1$, where n is the PRBS order, to test the interfaces.

Depending on the specifications of the protocol being implemented, the PRBS bit order needs to be selected accordingly. For example, Serial Digital Interface (SDI) for high-definition video transmission recommends having a length of sequence of 511, or $(2^9 - 1)$.

The timing information in Fig. 5.15 are:

- Mask—Blue area in the center of the eye diagram indicates minimum requirements of a particular standard or protocol. This is a keep-out area, where any waveform that touches the blue area is a timing violation.
- V_{ih}—Input high voltage. Minimum voltage for a HIGH logic level.
- V_{il}—Input low voltage. Minimum voltage for a LOW logic level.
- T_{su}—Setup time. Minimum setup time.
- T_{ho}—Hold time. Minimum hold time.
- Noise—Total noise, signal reflections, and power supply noise.
- Jitter—Total jitter, including any timing uncertainty and noise causing timing errors.

Analyzing all the timing information in the eye diagram is difficult to do manually, but there are tools on the market that can be leveraged to do the compliance tests automatically. For example, HyperLynx SERDES Wizards [3] have bundled many standard compliance tests like Ethernet, USB, and PCIe for designers to use. Chapter 10 shows the use of HyperLynx Wizards to do a USB 3.1 channel design.

5.3   Digital-to-Analog Converter (DAC)

A DAC converts the digital input codes to the analog output values and its transfer function is shown in Fig. 5.16 [1]. Equation (5.15) shows the relationship between the digital word, B_{in}, and the analog output signal, V_{out}.

$$V_{out} = V_{ref}\left(b_1 2^{-1} + b_2 2^{-2} + b_3 2^{-3} + \ldots + b_N 2^{-N}\right) = V_{ref}B_{in}, \qquad (5.15)$$

where

$$B_{in} = b_1 2^{-1} + b_2 2^{-2} + b_3 2^{-3} + \cdots + b_N 2^{-N}.$$

Similar to ADC, for the 3-bit DAC and V_{ref} of 8 V, the V_{LSB} is equal to 1 V. Table 5.2 shows an example of a 3-bit DAC taking a digital word and converting it to an equivalent analog level assuming that the sampling error is ±0.5 LSB.

Figure 5.17 shows a practical DAC block diagram where the digital input is being converted to analog and the Reconstruction Filter eliminates the sampling noise modulated on the analog waveform. The design of this filter depends on which DAC is being used and how much noise suppression is necessary to achieve a certain signal-to-noise specification. The filter topologies and design methodologies are covered in Chap. 4.

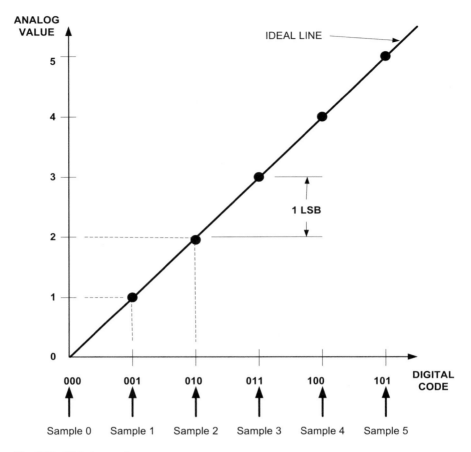

Fig. 5.16 Digital-to-analog

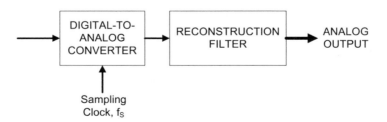

Fig. 5.17 Practical digital-to-analog

The filter requirements for the DAC are analog bandwidth, f_A, samples per second input, f_s, and stop band attenuation. As shown in Fig. 5.4, the sampled data has an image closest to the band of interest at

Table 5.2 DAC input and output

Digital input code	Analog output (V)	Error (V)
000	0	0.5
001	1	±0.5
010	2	±0.5
011	3	±0.5
100	4	±0.5
101	5	±0.5
110	6	±0.5
111	7	±0.5

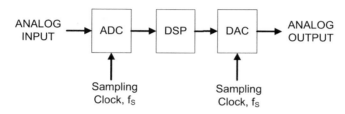

Fig. 5.18 DSP system block diagram

$$\text{Image} = f_s - f_A.$$

For example, a video signal has a bandwidth of 6 MHz, and the DAC input is 27 MSPS (Mega Samples Per Second). The image of this video signal is

$$\text{Image} = 27\ \text{MHz} - 6\ \text{MHz} = 21\ \text{MHz}.$$

If the system needs a 60 dB signal-to-noise performance, then the image needs to be attenuated at least 60 dB at 21 MHz. The details of the filter design are in Chap. 4.

5.4   Practical Data Converter Design Considerations

A typical high-level block diagram shown in Fig. 5.18 consists of input ADC [4], DSP, and DAC [5]. Here are the important parameters to consider when selecting ADCs and DACs:

- Resolution and signal-to-noise (SNR)
- Input and output voltage range
- Sampling frequency
- Differential nonlinearity
- Integral nonlinearity

5.4.1 Resolution and Signal-to-Noise

ADC or DAC resolution determines the number of digital bits to represent an analog signal. Higher resolution always yields higher SNR. Here is an SNR specification of ADC [4].

For a 10-bit ADC, SNR = 55 dB for f_A = 10 MHz and f_s = 110 MSPS.

Let us calculate the Effective Number of Bits or ENOB using Eq. (5.9).

$$\text{ENOB} = \frac{\text{SNR} - 1.76}{6.02} = \frac{55 - 1.76}{6.02} \cong 9.$$

ENOB indicates that the performance of this 10-bit ADC is equivalent to the performance of an ideal 9-bit ADC. Achieving the theoretical resolution is challenging, so getting 9-bit performance out of a 10-bit ADC is considered a very good ADC.

5.4.2 Sampling Frequency

Sampling frequency determines the frequency spacing between the analog signal and its images as shown in Fig. 5.4. So, a higher sampling frequency is preferred. This allows lower order input and output filters. The disadvantage of higher sampling frequency is that it tends to radiate more effectively as more traces on a PCB become effective antennas at higher frequency. Refer to Chap. 15 for more details on EMI.

In the ADC [4] data sheet, the SNR specification shows a sampling frequency of 110 MHz for a 10 MHz signal. The minimum sampling frequency is

$$\text{At Nyquist,} \quad f_s = 2f_A = 2(10 \text{ MHz}) = 20 \text{ MHz},$$
$$\text{Oversampling} = 110 \text{ MHz}/20 \text{ MHz} = 5.5.$$

The oversampling ratio of 5.5 indicates that the ADC samples the input signal at a rate 5.5 times higher than the minimum sampling frequency required by the Nyquist Theorem. In general, oversampling ADC or DAC has higher performance than the non-oversampling one.

5.4.3 Input and Output Voltage Range

ADC and DAC datasheets specify a maximum input voltage and a minimum output voltage, respectively. To maximize the performance of the system, it is recommended to amplify or attenuate the signal to get the maximum symmetrical

Fig. 5.19 Inputs to ADC

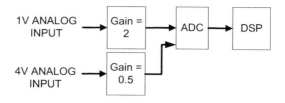

Fig. 5.20 Voltage divider

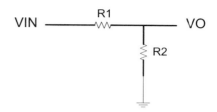

swing. For example, an ADC has two inputs, and each input has a maximum input specification of 2 V peak-to-peak. The two outputs from the previous stages measure 1 and 4 V peak-to-peak. To balance the ADC inputs, one input path must include a gain circuit and the other has an attenuator as shown in Fig. 5.19.

The gain stages of 2 and 0.5 can be implemented using op amps as shown in Chap. 4. Another option to design a gain of 0.5 is using a resistor divider to divide the input voltage by half. Here is an example.

For the voltage divider shown in Fig. 5.20, the output voltage is

$$V_{\mathrm{o}} = \frac{R_2}{R_1 + R_2} V_{\mathrm{IN}}. \tag{5.16}$$

If $R_2 = R_1$, then V_{o} is half of V_{IN}.

Regarding the gain circuit of 2, refer to Chap. 4 for more details.

Similarly, for the DAC output voltage range, if the DAC being used has a low-level voltage output that is not compatible with the next stage or not compliant with some input and output standards, then a gain circuit at the DAC output is required. Refer to Chap. 4 to design this amplifier circuit.

5.4.4 Differential Nonlinearity (DNL)

The Differential Nonlinearity (DNL) error [6] occurs on both ADC and DAC and is defined as the difference between the step width and one LSB. From Fig. 5.21 [6], the ADC DNL errors occur at 1 and 4 V analog levels. Ideally, the 001 code is centered at 1 V with $\pm 1/2$ LSB tolerance and the 100 code is centered at 4 V with $\pm 1/2$ LSB, but due to DNL errors, the 001 code is at 1 V with $-1/2$ LSB tolerance and the 100 code is at 4 V with $-1/2$ LSB to $+1$ LSB tolerance.

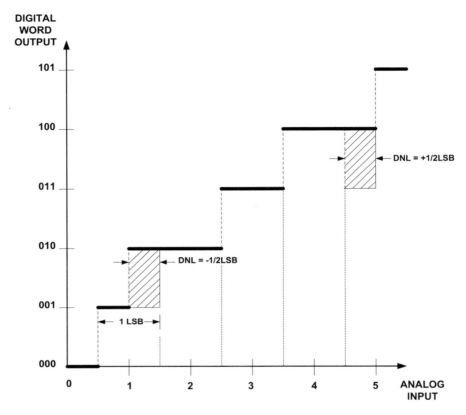

Fig. 5.21 ADC differential nonlinearity

Figure 5.22 [6] shows a DAC DNL error where the input 010 code has an analog range larger than 1 LSB voltage. In this case, the error is +1/4 LSB.

5.4.5 *Integral Nonlinearity (INL)*

The Integral Nonlinearly (INL) is defined as the maximum deviation of the actual transfer function from an ideal straight line. Like DNL, both ADC and DAC have INL errors. For ADC, the amount of error is measured at the transition from one digital code to another and compared it to the ideal transition point as shown in Fig. 5.23 [6]. For example, the actual transition from 001 code to 010 code happens at the 1 V point instead of 1.5 V point. This early transition produces a −1/2 LSB INL error.

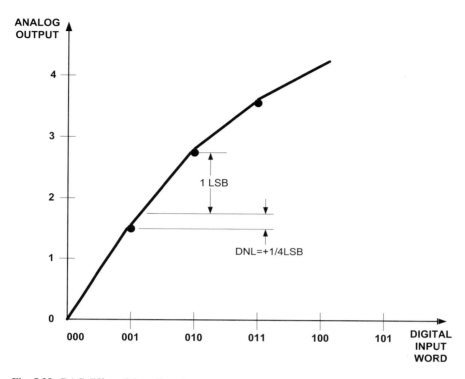

Fig. 5.22 DAC differential nonlinearity

Figure 5.24 [6] shows an INL error caused by the DAC. In this case, the maximum deviation from the ideal curve happens at the 011 input digital code and the error is equal to 1/2 LSB.

5.5 Summary

Important points to remember in this chapter are:

- Proper sampling and filtering are key to achieving high-performance data converter system design. Understanding Nyquist and oversampling techniques are crucial for making the right data converter selection for the targeted application.
- The derivation of SNR in this chapter assumed that there is only quantization noise in the system. This is an ideal case and is not practical in real-world design. ENOB is a good measurement of the actual data converter performance.
- Leveraging automated design tools to analyze eye diagram timing information and to run industry standard compliance tests such as PCIe, DDR, and USB.

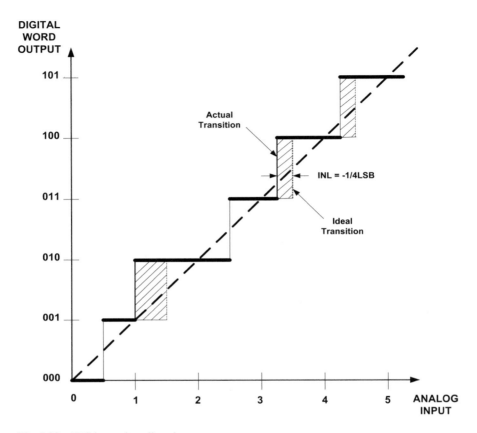

Fig. 5.23 ADC integral nonlinearity

- DNL and INL errors occurred in ADC and DAC cannot be eliminated using system design techniques. These are inherent in the data converter itself. Choose the converters with the lowest DNL and INL specifications if possible.
- Overall, the design goal for the data converter system is to match the performance specified in the converter datasheet. For example, if the converter has an 80 dB SNR specification, the system performance target should be 80 dB SNR. This is the best performance that can be achieved because all other components (anti-aliasing filters, amplifiers, power supplies, and reconstruction filters) around the converter tend to generate additional noise and errors in the system.

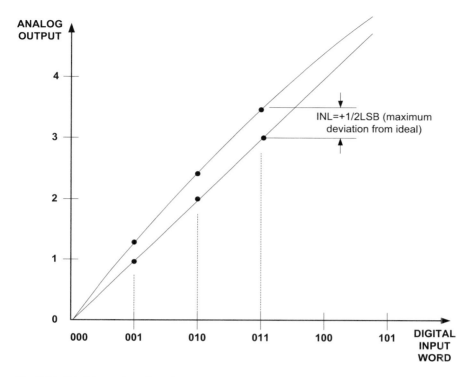

Fig. 5.24 DAC integral nonlinearity

References

1. D. Johns, K. Martin, *Analog Integrated Circuit Design* (John Wiley & Sons, New York, 1997)
2. Texas Instruments Inc., Interface Circuits for TIA/EIA-644 (LVDS) Design Notes, SLLA038B (2002)
3. Mentor Graphics (HyperLynx VX2.7) HyperLynx Signal Integrity Simulation Software. http://www.mentor.com/products/pcb-system-design/circuit-simulation/HyperLynx-signal-integrity/
4. Texas Instruments Inc., TVP7002 Triple 8-/10-Bit 165-/110-MSPS Video and Graphics Digitizer With Horizontal PLL, SLES205A (2007)
5. Texas Instruments Inc., THS8200 All-Format Oversampled Component Video/PC Graphics D/A System with Three 11-Bit DACs, CGMS Data Insertion, SLES032B (2009)
6. Texas Instruments Inc., Understanding Data Converters, Mixed-Signal Products, SLAA013 (1995)

Chapter 6
Transmission Line (TL) Effects

Transmission line (TL) effects are one of the most common causes of noise problems in high-speed DSP systems. When do traces become TLs and how do TLs affect the system performance? A rule-of-thumb is that traces become TLs when the signals on those traces have a rise time (T_r) less than twice the propagation delay (T_p). For example, if a delay from the source to the load is 2 nS, then any of the signals with a rise time less than 4 nS becomes a TL. In this case, termination is required to guarantee minimum overshoots and undershoots caused by reflections. Excessive TL reflections can cause electromagnetic interference and random logic or DSP false triggering. As a result of these effects, the design may fail to get the FCC certification or to fully function under all operating conditions such as at high temperatures or over-voltage conditions.

There are two types of transmission lines, lossless and lossy. The ideal lossless transmission line has zero resistance while a lossy TL has some small series resistance that distorts and attenuates the propagating signals. In practice, all TLs are lossy. Modeling of lossy TLs is a difficult challenge that is beyond the scope of this book. Since the focus of this book is only on practical problem-solving methods, it assumes a lossless TL to keep things simple. This is a reasonable assumption because in DSP systems where the operating frequency is less than 1 GHz the losses on printed circuit board traces are negligible compared to losses in the entire signal chain, from analog to digital and back to analog.

6.1 Transmission Line Theory

A lossless TL is formed by a signal propagating on a trace that consists of a series of parasitic inductors and parallel capacitors as shown in Fig. 6.1.

The speed of the signal, V_p, is dependent on properties such as characteristic impedance, Z_o, which is defined as an initial voltage V^+ divided by the initial current I^+ at some instant of time. Equations (6.1) and (6.2) for V_p and Z_o are

© The Author(s), under exclusive license to Springer Nature Switzerland AG 2023   81
T. T. Tran, *High-Speed System and Analog Input/Output Design*,
https://doi.org/10.1007/978-3-031-04954-5_6

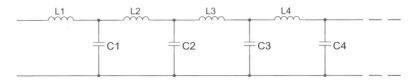

Fig. 6.1 Lossless transmission line model

$$V_p = \frac{1}{\sqrt{LC}}, \qquad (6.1)$$

$$Z_o = \sqrt{\frac{L}{C}}, \qquad (6.2)$$

where L is inductance per unit length and C is capacitance per unit length.

Another important property of the TL is the propagation delay, T_d. Equation (6.3) for T_d is

$$T_d = \frac{1}{V_p} = \sqrt{LC}. \qquad (6.3)$$

The source and load TL reflections depend on how well the output impedance and the load impedance, respectively, are matched with the characteristic impedance. The load and source reflection coefficients, Eqs. (6.4) and (6.5), are

$$\Gamma_S = \text{source\_reflections} = \frac{Z_S - Z_O}{Z_S + Z_O}, \qquad (6.4)$$

$$\Gamma_L = \text{load\_reflections} = \frac{Z_L - Z_O}{Z_L + Z_O}, \qquad (6.5)$$

where Z_S and Z_L are the source impedance and load impedance, respectively.

The following example shows the characteristics of a TL with no load and with a 3 V signal source driving the line.

Example 6.1 Calculate the voltage at the open-ended load of the transmission line below (Fig. 6.2).

$$V_{\text{initial}} = V_{\text{clk}} \frac{Z_O}{Z_S + Z_O} = 3 \frac{50}{25 + 50} = 2 \text{ V},$$

$$\rho_S = \frac{Z_S - Z_O}{Z_S + Z_O} = \frac{25 - 50}{25 + 50} = -0.333,$$

$$\rho_L = \frac{Z_L - Z_O}{Z_L + Z_O} = 1.$$

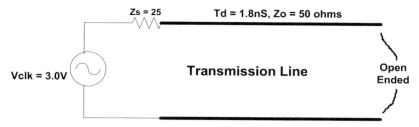

Fig. 6.2 Open-ended transmission line

Fig. 6.3 Lattice diagram of
open-ended TL

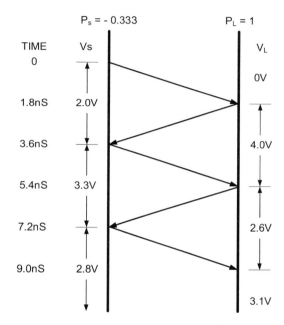

In Fig. 6.3, the overshoot voltage can be calculated using a lattice diagram [1] as follows (Fig. 6.4).

At $T1 = 1.8$ nS: $V_L = V_i + \rho_L V_i = 2 + 2 = 4.0$ V,
At $T2 = 3.6$ nS: $V_S = 4.0$ V $+ \rho_S(\rho_L V_i) = 4.0$ V $- 0.67$ V $= 3.33$ V,
At $T3 = 5.4$ nS: $V_L = 3.33$ V $+ \rho_L(\rho_S(\rho_L V_i)) = 3.33 - 0.67 = 2.66$ V,
At $T4 = 7.2$ nS: $V_S = 2.66$ V $+ \rho_S[\rho_L(\rho_S(\rho_L V_i)] = 2.66 + 0.22 = 2.88$ V,
At $T5 = 9.0$ nS: $V_L = 2.88 + \rho_L\{\rho_S[\rho_L(\rho_S(\rho_L V_i)]\} = 3.1$ V.

As shown in Example 6.1, the reflections with a 3 V source caused the signal to overshoot as high as 4 V at the load as explained below:

- The initial voltage level at the load at time $T1$ depends on the load impedance, which is infinite for an open load, and the characteristic impedance of the TL.
- The voltage level at time $T2$, when the reflected signal arrives at the source, depends on the source impedance and the characteristic impedance of the TL.

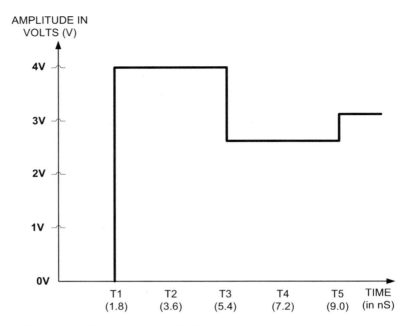

Fig. 6.4 Voltage waveform at the open-ended load

- The voltage level at time $T3$, when the reflected signal arrives at the load again, depends on the reflected voltage at $T2$ plus the reflected voltage at time $T3$.
- This process continues until a steady state is reached. In this example, the steady state occurs at $T5$, which is 9 nS from $T1$, as shown in Fig. 6.3.

Figure 6.5 shows the waveforms at the load for both terminated and unterminated circuits. As shown in the previous example, the terminated TL has a zero-reflection coefficient and therefore no ringing occurs on the waveform as seen on the top graph of Fig. 6.5. The problem is that in high-speed digital design, adding a 50-Ω resistor to ground at the load is not practical because this requires the buffer to drive too much current per line. In this case, the current would be 3.3 V/50 = 66 mA. A technique known as parallel termination can be used to overcome this problem. It consists of adding a small capacitor in series with the resistor at the load to block DC. The RC combination should be much less than the rise and fall times of the signal propagating on the trace.

Figure 6.6 shows a parallel termination technique. This method can be used in the application where one output drives multiple loads as long as the traces to the loads called L_2 are a lot shorter than the main trace L_1.

To use the parallel termination technique, it is necessary to calculate the maximum allowable value for L_2 according to Eq. (6.6) below assuming the main trace L_1 and the rise time T_r are known.

$$L_2, \max = L_1^{\frac{T_r}{10}}. \tag{6.6}$$

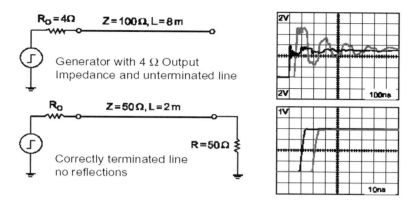

Fig. 6.5 Voltage waveforms at the terminated and unterminated loads

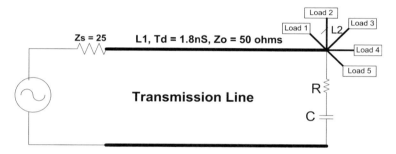

Fig. 6.6 Parallel termination with multiple loads

6.2 Parallel Termination Simulations

Parallel termination techniques become useful when designers must use a single clock output to drive multiple loads to minimize the clock skew between the loads. In this case, having a series resistor at the source limits the drive current to the loads and may cause timing violations by increasing rise times and fall times. This simulation example includes one 6 in. trace (L_1) and two 2 in. stubs. The DSP [2] drives the main L_1 trace and one memory device connected to each end of the 2 in. trace. It is reasonable to neglect the effects of the stubs provided they are short and meet the criteria shown in Fig. 6.7. In this case, only one parallel termination (68 Ω and 10 pF) is required at the split of the main trace to the loads. Referring to the simulation result in Fig. 6.8, the waveforms at the loads look good and meet all the timing requirements for the memory devices. As expected for the "no series" termination case, the waveform at the source does not look good but this does not affect the system integrity at the load.

Figure 6.9 shows an example of one clock output driving two loads connected using a daisy-chain topology. The distance from the source to the first load (first

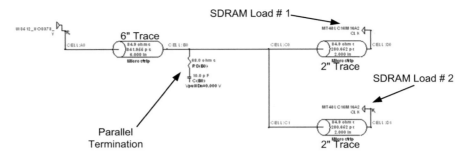

Fig. 6.7 Parallel termination configuration

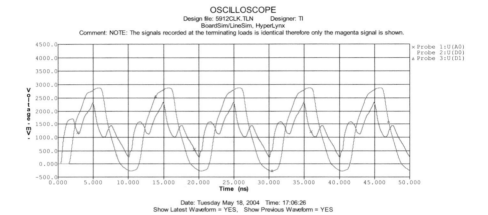

Fig. 6.8 Parallel termination simulation results

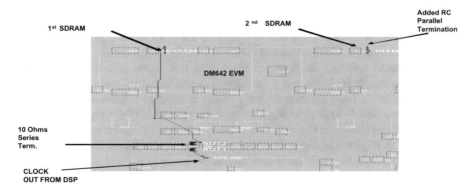

Fig. 6.9 DSP clock driving two loads

SDRAM) is the same as the distance from the first load to the second load (second SDRAM). In this case, the reflections coming from the second load distort the clock signal at the first load. The best way to minimize this distortion is by adding a parallel termination at the second load to reduce the impedance mismatch and therefore reduce the reflections as shown in Figs. 6.10 and 6.11. This system still requires a

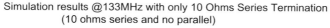

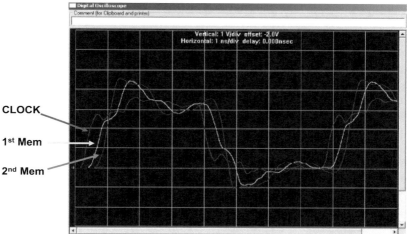

Fig. 6.10 Clock distortion due to reflections

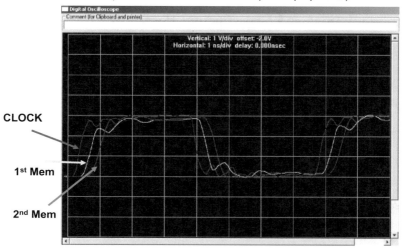

Fig. 6.11 Clock waveform with a parallel termination

series termination at the source to control the edge rate of the whole signal trace. This resistor needs to be small so that the source and sink currents are large enough to drive two loads. In this example, the series termination resistor is 10 Ω.

6.3   Practical Considerations of TL

In general, high-speed DSP systems consist of many CMOS devices where the input impedance is very high, typically in MΩ and the input capacitance is relatively small, less than 20 pF. In this case, with no load termination, the TL looks like a transmission line with a capacitive load, rather than an open circuit. The capacitive load helps reduce the rise time and allows the designers to use only a series termination at the source. This approach is becoming quite common in high-speed systems.

In Fig.. 6.12, the voltage at the load is slowly charged up to the maximum amplitude of the clock signal. Initially, the load looks like a short circuit. Once the capacitor is fully charged, the load becomes an open circuit. The source resistor Z_S controls the rise and fall times. Higher source resistance yields slower rise time. The load voltage at any instant of time, t, greater than the propagation delay time, can be calculated using the following equation:

$$V_L = V_{clk}\left(1 - e^{-(t-T_d)/\tau}\right),\tag{6.7}$$

where t is some instant of time greater than the propagation delay and $\tau = C_L Z_o$, where C_L and Z_o are the load capacitor and characteristic impedance, respectively.

6.4   Simulations and Experimental Results of TL

6.4.1   TL Without Load or Source Termination

One of the well-known techniques to analyze the PC board is using a signal integrity software [3] to simulate the lines. Figure 6.13 shows a setup used for the simulations.

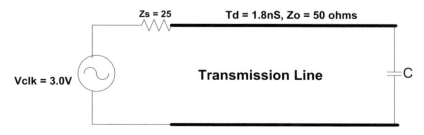

Fig. 6.12 Practical model of TL

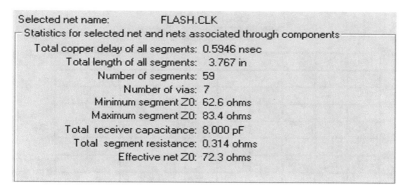

Fig. 6.13 Simulation setup

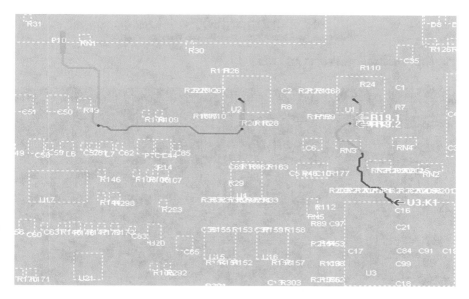

Fig. 6.14 PC board showing FLASH.CLK trace

The selected signal is FLASH.CLK which is a clock signal generated by a DSP. Figure 6.14 shows an actual PC board designed with a DSP where the clock is driven by U3 and is measured at U2.

Figure 6.15 shows the simulation result at U2, and Fig. 6.16 shows the actual scope measurement in the lab.

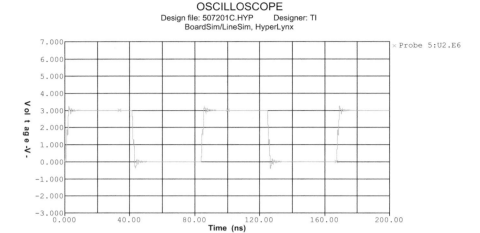

Fig. 6.15 Simulation result of FLASH.CLK

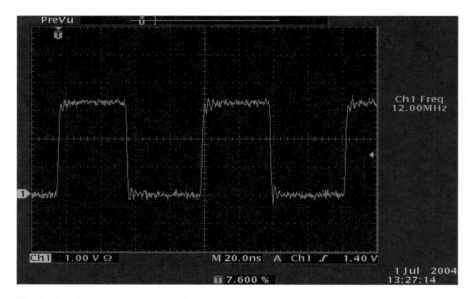

Fig. 6.16 Lab measurement of FLASH.CLK

6.4.2   TL with Series Source Termination

As discussed earlier, most high-speed system designs use this technique since it is possible to optimize the load waveforms simply by adjusting the series termination resistors. This technique also helps reduce the dynamic power dissipation, since the initial drive current is limited to the maximum source voltage divided by the characteristic impedance. Figure 6.17 shows the setup used for the simulation of the audio clock driven by an audio CODEC external to the DSP.

Figure 6.18 shows an audio clock that transmits by U17 and receives by U3. The design has a 20-Ω series termination resistor but no parallel termination at the load. This demonstrates the concept discussed earlier.

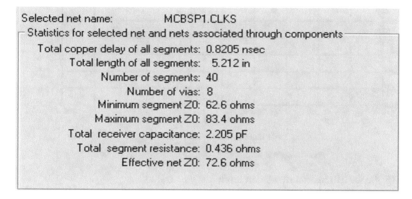

Fig. 6.17   Series termination clock setup

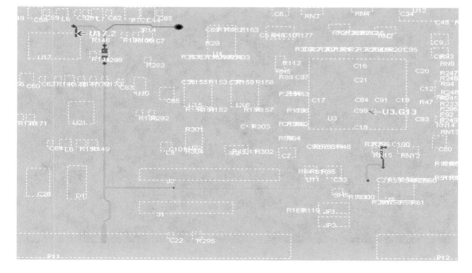

Fig. 6.18   Audio clock with series termination

The simulation result is shown in Fig. 6.19.

The lab measurement shown in Fig. 6.20 correlates with the simulation very well. The 22-Ω series resistor can be modified to lower the overshoots and undershoots. But since the overshoots are less than 0.5 V, they are acceptable in this case.

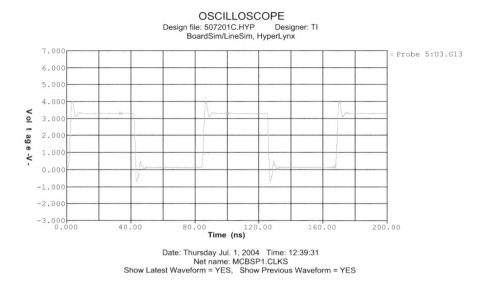

Fig. 6.19 Series termination simulation result

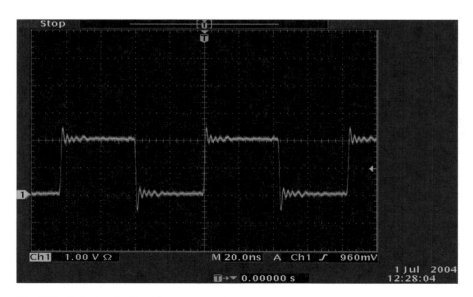

Fig. 6.20 Series termination lab measurement

6.5 Ground Grid Effects on TL

In summary, the simulation results correlate very well with the actual lab measurements. Designers need to understand the TL characteristics and terminate traces to minimize reflections that may cause random circuit failures, excessive noise injected into the power, ground planes, and electromagnetic radiation.

One final comment about the TL is that the previous examples were based on a model where a signal trace is on top of a ground plane known as a microstrip model. Other techniques, such as a ground grid, are also commonly used. Example 6.2 demonstrates the effects of the ground grid. In this configuration, the designers need to understand the current flows and their effect on the characteristic impedance.

Example 6.2 Figure 6.21 shows an example of using a ground grid, instead of ground plane, for the PC board. As shown in this figure, the current path is not immediately under the signal trace, so there is a large current return loop that yields higher inductance and lower capacitance per unit length. In this case, the characteristic impedance is higher than if a continuous ground plane was used.

Figure 6.22 also shows another example of using a ground grid where the signal is being routed diagonally. As shown in this figure, the current return has to travel on a zig-zag pattern back to the source and creates a large current return loop that yields higher inductance and lower capacitance per unit length. In this case, the characteristic impedance is higher than using a continuous ground plane and higher than the case where the signal is routed in parallel with the ground grid as shown in Fig. 6.21.

So, if ground grid is required in a design, the best approach is to route the high-speed signals right on top of the grids and parallel to the grid to ensure the smallest current return loops. This lowers the characteristic impedance to the level equivalent to the impedance of the continuous ground plane. This is difficult to accomplish since a complex board has many high-speed traces. Therefore, continuous ground plane is still the best method to keep characteristic impedance and EMI low.

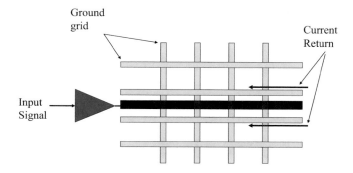

Signal trace is routed between the two ground paths of the grid

Fig. 6.21 Current return paths of ground grid

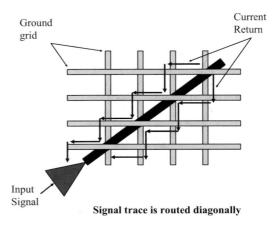

Fig. 6.22 Current return paths of diagonal grid

Ground grid

Current Return

Input Signal

Signal trace is routed diagonally

6.6 Minimizing TL Effects

As demonstrated in this chapter, transmission line effects cause signal distortions which may lead to digital logic failures and radiations. These effects cannot be eliminated totally but can be minimized by applying the following guidelines:

- Slow down the signal edge rate by lowering the buffer drive strength if it is not affecting the timing margins. Remember a trace becomes a TL if the rise time of the signal propagating on it is less than a roundtrip propagation delay.
- If the edge rate of the buffer is not configurable, add a series termination resistor as close to the source as possible. The value of the resistor is equal to the characteristics impedance (Z_o) of the trace minus the buffer output impedance.
- When one clock is driving multiple loads as shown in Fig. 6.9, add a series termination at the source and a parallel termination at the load.
- If ground grids must be used, route the high-speed signals in parallel with the ground grids to reduce the current return loops. Ground planes are always preferred over ground grids.

References

1. S. Hall et al., *High Speed Digital System Design* (John Wiley & Sons, New York, 2000)
2. Texas Instruments Inc., OMAP5912 Applications Processor Data Manual (2003). http://focus.ti. com/lit/ds/symlink/omap5912.pdf
3. Mentor Graphics, Hyperlynx Signal Integrity Simulation Software (2004). http://www.mentor. com/products/pcb-system-design/circuit-simulation/hyperlynx-signal-integrity/

Chapter 7
Transmission Line (TL) Effects in Frequency Domain

Transmission line (TL) effects are one of the most common causes of noise and design problems in high-speed systems. Traditionally, in digital design, the switching frequency and the rise or fall time of the signal could generate excessive overshoots or undershoots which can cause timing violations as described in Chap. 6. As the operating frequency gets higher, >100 MHz, the digital waveform must change to accommodate for fast switching, lower supply voltage, and lower rise and fall times. The trend makes it more difficult for designers to control noise and radiation while maintaining good signal and power integrities.

One of the most effective ways to reduce noise, EMI, and PCB routing area is reducing the number of inputs and outputs to and from an IC. The common method for device interconnects is converting parallel busses to serial busses in a system. But to be logically compatible with parallel busses, the serial busses must run at a much higher speed, depending on how many parallel data bits are being serialized. For example, an 8-bit parallel bus running at 100 MHz can be directly converted to a single-bit serial bus running at 800 MHz, the number of parallel bits multiplied by the switching speed. Converting parallel-to-serial and serial-to-parallel method is known as SERDES, Serializer, and Deserializer.

To push the switching speed to multi-gigahertz range, up to 32 Gbps, the industry has adapted modeling and simulating techniques commonly being used in microwave circuits, known as scattering parameter or *s*-parameter. This chapter covers the fundamentals of *s*-parameter and how it is being applied in a high-speed digital system.

7.1 *S*-Parameter Fundamentals

With data rate running up to multi-gigahertz speed, the traditional method of digital design is no longer valid. Digital designers must learn and adapt RF/microwave design techniques in modeling systems or components using scattering parameters

© The Author(s), under exclusive license to Springer Nature Switzerland AG 2023   95
T. T. Tran, *High-Speed System and Analog Input/Output Design*,
https://doi.org/10.1007/978-3-031-04954-5_7

or *s*-parameters. The basic form of *s*-parameter model is a two-port network shown in Fig. 7.1. The *s*-parameters, *S*11, *S*12, *S*21, and *S*22, represent the reflection and transmission coefficients of the network.

This network has two ports, Port 1 and Port 2, and the parameters of the ports are as follows:

- *S*11, Return loss—Reflection coefficient in which the voltage applied to Port 1 is reflected back to Port 1.
- *S*12, Insertion loss—Transmission coefficient in which the voltage applied to Port 2 is transferred to Port 1.
- *S*21, Insertion loss—Transmission coefficient in which the voltage applied to Port 1 is transferred to Port 2.
- *S*22, Return loss—Reflection coefficient in which the voltage applied to Port 2 is reflected back to Port 2.

In *s*-parameters, the network is assumed to have proper termination at the output, and the unit of *s*-parameter is in decibel, dB, output voltage divided by input voltage as in Eq. (7.1).

$$S_{ij} = \frac{\text{Output Voltage, } j}{\text{Input Voltage, } i}, \tag{7.1}$$

$$\text{Mag}(s), \text{Magnitude} = \frac{\text{Output Voltage Amplitude}}{\text{Input Voltage Amplitude}}, \tag{7.2}$$

$$S_{\text{dB}} = 20 \, \log_{10}(\text{Mag}(s)), \tag{7.3}$$

$$\text{Phase}(s) = \text{Phase(output)} - \text{Phase(input)}. \tag{7.4}$$

The most popular network used for analyzing differential pairs is a four-port network shown in Fig. 7.2. In the four-port network, the naming convention of the ports is the same as the two-port network in which *S*12 is a transmission coefficient when a voltage applied to Port 2 is transferred to Port 1.

In SERDES protocol, before transmitting a single-ended signal from one device or system to another, the single-ended signal is first converted to one different pair,

Fig. 7.1 Two-port network

Fig. 7.2 Four-port network

which has a positive component and a negative (180° out of phase) component of the original signal. Then, the differential pair is transmitted on a PCB to its destination. The advantages of differential signaling are good noise immunity, low radiation, and low dynamic power consumption as described in Chap. 5.

In Fig. 7.2 four-port network, the insertion losses (IL) are *S*12, *S*21, *S*34, and *S*43. And the return losses (RL) are *S*11, *S*22, *S*33, and *S*44. In addition to insertion and return losses, the four-port network has crosstalk terms, *S*13, *S*31, *S*14, *S*41, *S*32, *S*23, *S*42, and *S*24. The crosstalk term is defined as the voltage applied at one port is coupled to another port. NEXT is a near-end crosstalk where a voltage applied to one port couples to the nearest port, and FEXT is a far-end crosstalk where a voltage applied to one port couples to the farthest port, not on the same channel as the transmitted port. For example, *S*14 is a far-end crosstalk when a voltage applied to Port 4 (far end port) couples to Port 1; Port 1 is not on the same transmission path as Port 4 (Fig. 7.3).

The four-port network *s*-parameters consist of different modes, known as mixed-mode *s*-parameters. The modes include how the network responds to differential and common mode stimulus signals. The table in Fig. 7.4 demonstrates different modes of the four-port mix-mode *s*-parameters.

Fig. 7.3 Four-port *S*-parameter matrix

S11 (RL)	S12 (IL)	S13 (NEXT)	S14 (FEXT)
S21 (IL)	S22 (RL)	S23 (FEXT)	S24 (NEXT)
S31 (NEXT)	S32 (FEXT)	S33 (RL)	S34 (IL)
S41 (FEXT)	S42 (NEXT)	S43 (IL)	S44 (RL)

			Stimulus			
			Differential		Common Mode	
			Port 1	Port 2	Port 1	Port 2
Response	Differential	Port 1	Sdd11	Sdd12	Sdc11	Sdc12
		Port 2	Sdd21	Sdd22	Sdc21	Sdc22
	Common Mode	Port 1	Scd11	Scd12	Scc11	Scc12
		Port 2	Scd21	Scd22	Scc21	Scc22

Fig. 7.4 Four-port mixed-mode *S*-parameters table

Stimulus: Differential Input; Port 1 and Port 2; Response: Differential Output; Port 1 and Port 2; Sdd11: Differential to Differential Return Loss; Sdd12: Differential to Differential Insertion Loss; Sdd21: Differential to Differential Insertion Loss, same as Sdd12; Sdd22: Differential to Differential Return Loss, same as Sdd11

Figure 7.4 shows Stimulus or input signals and Response or output signals. And each input pair or output pair could be Differential Mode or Common Mode. For example, Quadrant 1 (upper left) of Fig. 7.4 is the most important quadrant and is the actual mode of operation for high-speed SERDES channels such as Gigabit Ethernet, USB 3.x, and PCIe. This quadrant defines the frequency response of the system or device.

Quadrant 2 (upper right, common mode to differential) and Quadrant 3 (lower left, differential to common mode) are cross-mode quadrants in which the parameters in Quadrant 2 show how well the system is designed to immune to radiation and in Quadrant 3 show how bad the system is radiating noise that could affect the performance.

Quadrant 4 parameters show how the system is rejecting common mode noise.

Design Example 7.1: Two-Port Network Design and Simulation Figure 7.5 shows a design of a two-port network, a 5.8 in. 50-Ω microstrip on a four-layer PCB.

HyperLynx BoardSim [1] simulation results in Fig. 7.6 show both Insertion Loss and Return Loss responses. For design optimization, it is preferable to make IL as small as possible to maintain good signal integrity while keeping RL large to ensure minimum reflected noise. In this design example, the IL is −6 dB at 10 GHz while the RL is −46 dB at the same frequency.

Design Example 7.2: Four-Port Network Design and Simulation Figure 7.7 shows a design of a four-port network, a 5.8 in. 50-Ω differential microstrip on a four-layer PCB, using the same PCB stackup as in Example 7.1.

Simulation results in Fig. 7.8 show both Insertion Loss and Return Loss responses. In this four-port network, the graphs in Fig. 7.8 were derived from parameters in Quadrant 1, differential stimulus, and differential response. Since all the ports were placed farther away from each other, the crosstalk terms in Quadrants 2 and 3 were insignificant, −200 dB down.

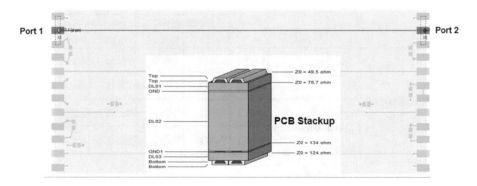

Fig. 7.5 Two-port network with PCB stackup

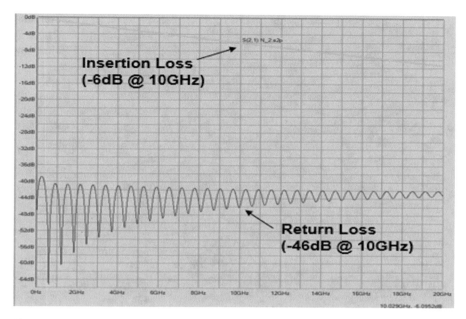

Fig. 7.6   HyperLynx simulations of two-port network

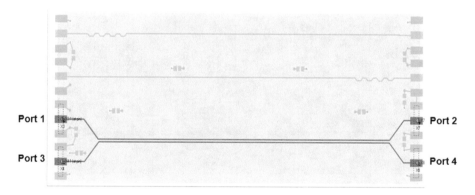

Fig. 7.7   HyperLynx BoardSim of four-port network

**Design   Example   7.3:   Four-Port   Network   Design   with   Retimer/
Redriver**   Figure 7.9 shows a design of a four-port network, a 5.8 in. 50-Ω differential
microstrip on a four-layer PCB, using the same PCB stackup as in Example 7.1.

In this example, there is a loss associated with PCB when transmitting high-speed
signal from Port 1 to Port 2. Whether or not the loss is acceptable depends on which
protocol is the design being implemented. If the channel loss causes the design to be
noncompliant with a certain industry standard, then the channel can be compensated

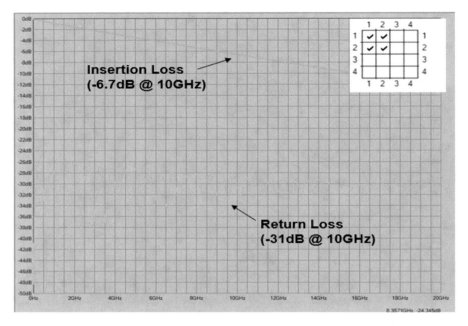

Fig. 7.8 HyperLynx simulations of four-port network

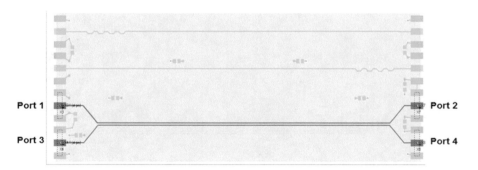

Fig. 7.9 HyperLynx BoardSim of four-port network

by inserting a Retimer or Redriver in the transmission to boost a particular frequency band of interest, similar to applying a channel equalizer.

To design with a Retimer, first designers need to obtain *s*-parameter models of the selected component from a manufacturer's website. Then place the model in the circuit shown in Fig. 7.10. Because *s*-parameter is a passive network modeling method, IC manufacturers provide multiple models for different gain curves in one device. For example, the device being used in this design is PI3EQX1001 Redriver [2] has many gain and equalizer curves, and each curve has one *s*-parameter model as shown below.

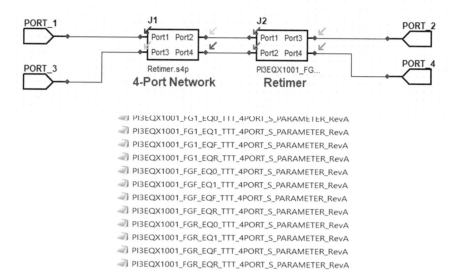

Fig. 7.10 HyperLynx LineSim circuit

The end-to-end HyperLynx LineSim [3] circuit shown in Fig. 7.10 consists of a four-port network model extracted from the layout in Fig. 7.9 and the Pericom PI3EQX1001 Redriver model (PI3EQX1001_FG1_EQ1_TTT_4PORT_S_PARAMETER_REVA).

Simulation results in Fig. 7.11 show the insertion loss curves with and without Retimer. With the Retimer, there is a gain of +9 dB at 10 GHz as compared to a loss of −6 dB at the same frequency without the Retimer.

7.2 Minimizing TL Effects in Frequency Domain

As demonstrated in this chapter, transmission line effects at high frequency cause signal quality degradation which can lead to system design failures and excessive noise and radiations. And to mitigate the issues, designs need to be simulated carefully using *s*-parameter models and need to be compensated accordingly.

Recommended guidelines for gigahertz digital design are:

- Simulate all the high-speed channels and verify the insertion loss of the channels are within the allowable limits. If not, use a Retimer or Repeater to compensate for the loss.
- If possible, in the PCB layout, avoid transitioning high-speed traces from one layer to another layer. Vias cause impedance discontinuity which leads to excessive insertion and return losses. Follow the industry standards if applicable. Some

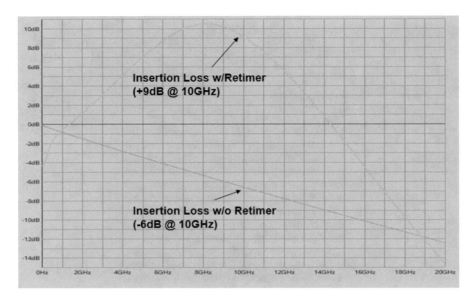

Fig. 7.11 Simulations of four-port network

standards only allow for two vias on a high-speed trace. Refer to Chap. 14 for more details about PCB layout,

* If vias are required for routing, make sure to add ground via next to the signal to allow for high-speed current return. Also, verify if the stubs on via cause any signal integrity issues or not. Methods like microvias or laser back-drilling are commonly being used to remove via stubs. Vias on high-speed traces must be modeled and included in the end-to-end simulations. Also, via placement is critical, so make sure to review the PCB guidelines required for the intended protocol. In general, place the via as close to the source or destination as possible.

References

1. Mentor Graphics HyperLynx Signal Integrity Simulation Software, BoardSim
2. Diodes Incorporated PI3EQX1001 Datasheet. https://www.diodes.com/assets/Datasheets/PI3 EQX1001.pdf
3. Mentor Graphics HyperLynx Signal Integrity Simulation Software, LineSim

Chapter 8
Effects of Crosstalk

In any electronic system, it is neither practical nor is it necessary to eliminate all the noise, as noise is not a problem until it interferes with the surrounding circuitries or radiates electromagnetic energy that exceeds FCC limits and or degrades the system performance. When noise interferes with other circuits it is called crosstalk. Crosstalk can be transmitted through electromagnetic radiation or electrically coupling, such as when unwanted signals propagate on the power and ground planes or couple onto the adjacent circuits. One of the most challenging problems designers are facing in today's electronic systems is to determine the source of crosstalk, especially in the case where crosstalk randomly causes system failures. Because components are so tightly packed into a very small printed circuit board (PCB). This chapter outlines the crosstalk mechanisms and design methodologies to minimize the effects of crosstalk.

8.1 Current Return Paths

In designing a system, it is crucial for designers to understand the current return paths as these current returns are the main sources of electromagnetical and electrical coupling. For example, the digital signal current return crosses the analog section of the design and causes noise on the analog waveforms or the current return generates a large current loop area, which radiates onto the adjacent circuitries.

Current returns follow different paths depending on their frequency. A high-frequency current flow tends to concentrate on the surface of the conductor as supposed to distribute uniformly across the conductor like a low-frequency current. This phenomenon is known as skin effect, and it modifies the current distribution and changes the resistance of the conductor. Due to skin effect, signals above 10 MHz tend to follow one return path while those below 10 MHz follow another. The low-speed signal current returns on the path of least resistance, normally the shortest route back to the source as shown in Fig. 8.2. In Fig. 8.1, the high-speed

T. T. Tran, *High-Speed System and Analog Input/Output Design*,
https://doi.org/10.1007/978-3-031-04954-5_8

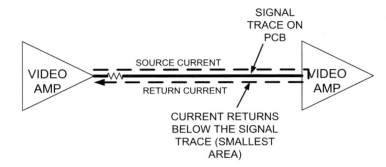

Fig. 8.1 High-frequency current return paths (>10 MHz)

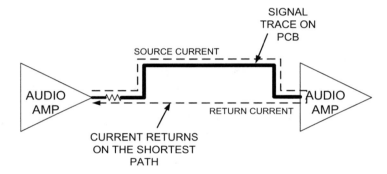

Fig. 8.2 Low-frequency current return paths (<10 MHz)

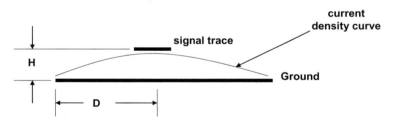

Fig. 8.3 Current return density

signal current, on the other hand, returns on the path of least inductance, normally underneath the signal trace. Knowing the current return paths is important for designers to optimize the system design to reduce crosstalk.

The current return density and the amount of crosstalk can be estimated as shown in Figs. 8.3 and 8.4. Based on the equations shown in the figures, the spacing between the traces and the distance that they run in parallel determines the amount of crosstalk. Obviously, moving the traces further from each other will reduce the crosstalk.

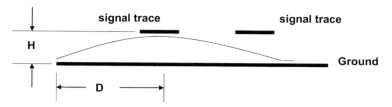

Fig. 8.4 Crosstalk estimation

The return current density, I_D, in Fig. 8.3 [1] is

$$I_D = \frac{I}{\pi H} \frac{1}{1 + \left(\frac{D}{H}\right)^2}, \tag{8.1}$$

where I is the signal current.

The crosstalk in Fig. 8.4 [1] can be estimated as

$$\text{Crosstalk} = \frac{K}{1 + \left[\frac{D}{H}\right]^2}, \tag{8.2}$$

where D is the distance between the traces, H is the height of the signal to the reference plane, and K is the coupling constant less than 1.

There are two types of crosstalk, forward and backward. Forward crosstalk, also known as capacitively coupled crosstalk. This occurs when the current flows in the same direction as the source. With backward crosstalk, which is also called inductively coupled crosstalk, the coupling current flows in the opposite direction of the source.

The following simulations [2] demonstrate the concept of reducing forward and backward crosstalk by spacing the aggressor and victim traces. The model simulates two parallel 5 mils wide, 12-in.-long traces. The source of the trace is connected to a DSP and the load to DDR memory. As shown in Fig. 8.5, $D0$ line is an aggressor and $D1$ line is a victim.

Figure 8.6 shows simulation results. On the victim trace, the first negative-going pulse, which has a -200 mV peak, is the forward crosstalk. The positive-going pulse of 240 mV is the backward crosstalk. The backward pulse width is about two times the coupling region. In this case, the coupling region is 3.54 nS and the simulation also shows a 4 nS backward crosstalk pulse.

The crosstalk Eq. (8.2) is

$$\frac{K}{1 + \left[\frac{D}{H}\right]^2}.$$

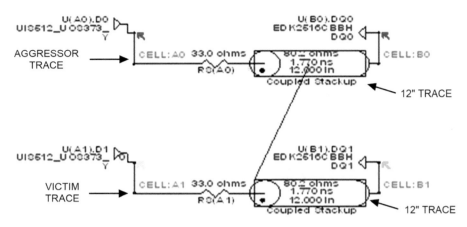

Fig. 8.5 Crosstalk simulation setup

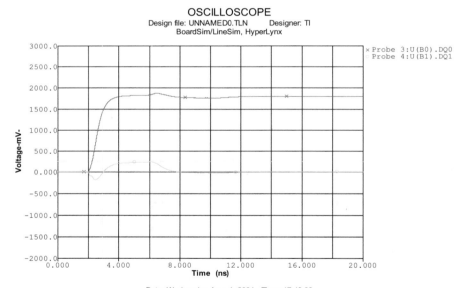

Fig. 8.6 Crosstalk simulation results for 5 mils spacing

Let us assume that $K = 1$, $D = 5$ mils, and $H = 10$ mils as in the simulation. The maximum crosstalk is then calculated as follows:

$$\text{Max. Crosstalk} = \frac{K}{1 + \left[\frac{D}{H}\right]^2} = \frac{1}{1 + \left[\frac{5}{10}\right]^2} = 0.8 \text{ V}.$$

As expected, the simulations in Fig. 8.6 showed that the peak-to-peak crosstalk is 440 mV, which is much less than the maximum crosstalk estimated.

Now, let us test the condition where the two traces are placed further away from each other, by making $D = 15$ mils. The maximum estimated crosstalk is now 0.3 V while the simulation in Fig. 8.7 shows a forward crosstalk of -100 mV and a backward crosstalk of 90 mV. The peak-to-peak crosstalk is about 190 mV, which is again much less than the calculated maximum of 300 mV. This simulation demonstrates how the rule-of-thumb provided earlier overestimates crosstalk.

In summary, accurately calculating and simulating the crosstalk of a system is not possible due to many complex capacitive and inductive coupling paths that are involved. The examples show how difficult it is to estimate and simulate crosstalk

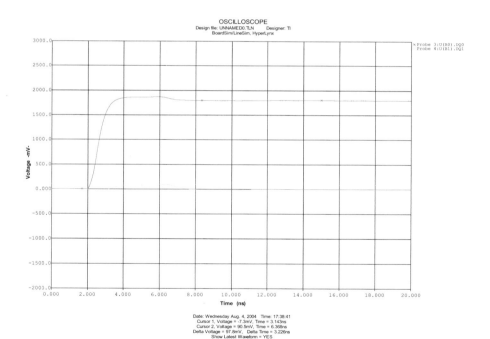

Fig. 8.7 Crosstalk simulation for 15 mils spacing

and the effects of spacing on the adjacent signal. The following points need to be considered before finalizing the design:

- When the PCB is designed, minimize the height, H, between the high-speed signal routing layer and the ground plane. Lower H yields lower crosstalk.
- Maximize the spacing, D, between the signals. Higher D yields lower crosstalk.
- For board layout, analyze the critical signals and minimize the coupling regions.
- Slow the edge rates if possible because this reduces crosstalk.

8.2   Crosstalk Caused by Radiation

Crosstalk can also be caused by high-speed signals that are routed on traces that form effective antennas. The first step in determining whether a trace is acting as an antenna is to calculate the wavelength of the signal using the following equation:

$$\lambda = \frac{C}{f}, \tag{8.3}$$

where C is the speed of light or 3×10^8 m/s and f is the frequency in Hz.

The equation shows that a 100 MHz clock signal has a wavelength of 3 m or 9.84 ft. A good rule for minimizing radiation is making sure that the trace length is not longer than the wavelength divided by 20. So, in the case of the 100 MHz clock signal, the signal length should be kept below 0.15 m or 0.492 ft. Keeping the traces below 0.5 ft is easy, but the square wave clock signal consists of multiple harmonics and each of the harmonics can radiate even when the traces are very short. Here is an example.

Example 8.1 Let $f = 500$ MHz, the fifth harmonic of the 100 MHz clock,

$$\lambda = \frac{C}{f} = \frac{3 \times 10^8}{500 \times 10^6} = 0.6 \text{ m}.$$

A rule-of-thumb for determining when the clock trace becomes an effective antenna is taking the wavelength divided by 20. The maximum length of the 500 MHz clock is

$$\lambda = \frac{C}{20f} = \frac{3 \times 10^8}{20(500 \times 10^6)} = 0.03 \text{ m or 3 cm which is 1.18 in.}$$

What this means is that depending on the rise and fall times of the 100 MHz signal, the energy of the fifth harmonic can radiate and interfere with the adjacent circuits when this signal trace is longer than 1.18 in. The energy of the harmonics depends on the rise and fall times of the signal as shown in Fig. 8.8. In this figure, it is assumed that the clock waveform has a 50% duty cycle, and rise and fall times are

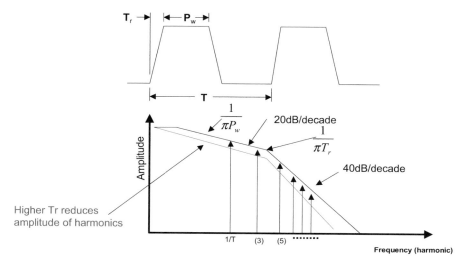

Fig. 8.8 Radiation caused by clock signal

equal. With these assumptions, only odd harmonics of the clock are present. The amplitude of the harmonics starts decaying at the first pole frequency, f_1, at a rate of 20 dB/decade and then increases to 40 dB/decade at the second pole frequency f_2. The equations for f_1 and f_2 are

$$f_1 = \frac{1}{P_w},$$ (8.4)

where P_w is the high time of the waveform.

$$f_2 = \frac{1}{\pi T_r},$$ (8.5)

where T_r is the rise time of the waveform.

To illustrate how a digital clock waveform can generate crosstalk that degrades the video quality of a time-shifting system [3], let us take a look at a system diagram shown in Fig. 8.9 where many critical components are placed on the same printed circuit board. In this design, the clock signals ranging from 18.4 to 100 MHz are being routed to all the subsystems (modem, audio CODEC, video encoder/decoder, CPU, MPEG encode, and MPEG-2 decode).

The time-shifting system in Fig. 8.9 operates as follows:

- As shown in Fig. 8.10, the Video Tuner receives a radio frequency (RF) signal from an antenna and demodulates the RF signal and converts it down to the baseband frequency.

Fig. 8.9 Time-shifting system

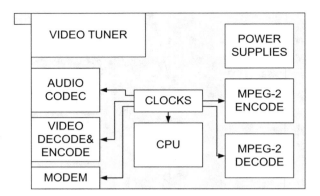

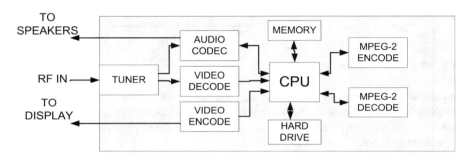

Fig. 8.10 Time-shifting system data flow

- The Video Decoder receives the baseband video signal from the Tuner and digitizes the signal to prepare for digital signal compression. This video data rate for an analog TV channel with $640 \times 480 \times 30$ frames/s resolution is around 147 Mbits/s (data rate $= 640 \times 480 \times 16$ bits/pixel $\times 30$ frames/s $= 147$ Mbits/s).

The CPU is responsible for running a high-level operating system and for managing all the video and audio data. The digital video data are captured by the CPU and are stored in external memories such as DDR.

The CPU transfers the digital video data stored in the memories to the MPEG-2 Encoder [4]. The Encoder compresses the data from 147 to 2 Mbits/s bit rate and sends the compressed bit rate back to the external memories. The compressed data are then sent to the hard drive for storage.

- The CPU then reads the compressed data from the hard drive and sends the data to the MPEG-2 Decoder [4]. The Decoder decompresses the data from 2 Mbits/s bit rate back to 147 Mbits/s bit rate and sends it to the Video Encoder. The Encoder converts the digital decoded data to analog signals and displays it on a TV screen.
- The Modem in the time-shifting system is there for communicating with the service provider server to request TV guides and software updates.

- The Audio CODEC is responsible for digitizing the analog audio signals received from the Tuner or an external audio device. The digital audio data follows the same path as the digital video signal in which the data are being compressed, stored, and playback the same way. This CODEC also receives the digital audio data from the CPU and converts it to analog signals for playing out to the speakers.

In this time-shifting system design, the crosstalk can occur anywhere within a system even though the system was fully FCC certified, so it is difficult to find the root causes of the problem. For example, any clock can generate harmonics that radiate to the antenna input and interfere with the TV channel. To illustrate this effect, Fig. 8.11 shows a video screen with horizontal lines generated by the third harmonic (55.2 MHz) of the 18.4 MHz clock radiating to the antenna input. 55.2 MHz harmonic happens to be within Channel 2 (54–60 MHz) of the NTSC spectrum [5] and causes interferences that cannot be rejected by the Tuner because the Tuner cannot distinguish between the in-band noise and the actual TV signal.

In this case, the best way to get rid of the interference is to reduce the energy radiated by the third harmonic of the 18.4 MHz clock. Figure 8.8 shows that increasing the rise time of the signal, assuming that this is not causing any setup and hold time violations, attenuates the harmonic amplitude and reduces the radiation. The two ways to reduce the rise time are lowering the clock buffer slew rate if possible or adding a series termination resistor at the output of the clock buffer as shown in Fig. 8.12.

It is very difficult to calculate the value of the resistor R, so the best method is to vary the resistance until the noise has longer appeared on the display. At this point,

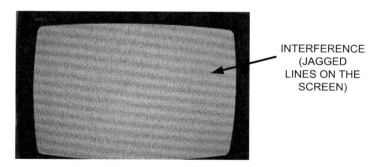

Fig. 8.11 Clock harmonic interfered with video

Fig. 8.12 18.4 MHz clock trace

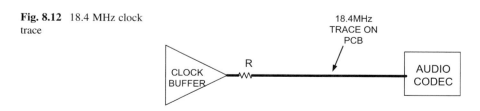

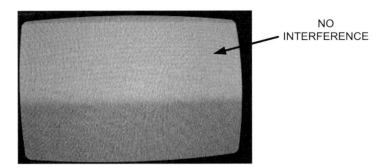

Fig. 8.13 Interference reduction by increasing the rise time

measure the rise and fall times of the 18.4 MHz clock and verify that reducing the rise and fall times does not cause any timing violations. Figure 8.13 shows a design example of having a 75-Ω series termination resistor to reduce the rise time and get rid of the interference as shown in Fig. 8.11.

8.3 Summary

As highlighted in Sects. 8.1 and 8.2, the current return paths and the signal rise time play the key role in generating crosstalk that interferes with the adjacent circuits and causes random system failures and or system performance degradation.

- Slow down the rise and fall times if possible. Increasing the rise time reduces the power spectral density of the harmonics as shown in Fig. 8.8.
- Keep high-speed signals as short as possible and make sure that the wavelengths of the third and fifth harmonics of the signal are much less than the wavelength divided by 20.
- Always add a series termination at the source of the clock signal as demonstrated in Fig. 8.12. This helps reduce the transmission line effects and provides a way to reduce the rise time if necessary.
- Always route the high-speed signals away from any critical high impedance traces. High impedance traces are the input traces to the video and audio amplifiers. Also, space the traces at least one width apart from each other to reduce the coupling; for example, a 5 mils trace should have a gap of at least 5 mils to another trace.
- The best method to minimize crosstalk is to place the clock generator component in the middle of the system as shown in Fig. 8.9. This ensures minimum routing clock traces to all other sections.

Using a spread spectrum clock generator is another way to reduce the peak radiated power but be careful with the jitter generated by these clock buffer devices. For example, in one spread spectrum clock buffer [6], the amplitude of the seventh harmonic can be attenuated by -13 dB by setting the spread spectrum at $\pm 2\%$. The issue with this setting is that the 100 MHz clock output jitter will increase by 529 ps. If this increase in jitter is still within the allowable limits of the DSP clock input, then this solution is acceptable.

References

1. H. Johnson, M. Graham, *High-Speed Digital System Design* (Prentice Hall, Englewood Cliffs, 1993)
2. Mentor Graphics, HyperLynx Signal Integrity Simulation Software (2004). http://www.mentor.com/products/pcb-system-design/circuit-simulation/hyperlynx-signal-integrity/
3. TIVO DVR. http://www.tivo.com/dvr-products/tivo-hd-dvr/index.html
4. International Organization for Standardization, Information Technology—Generic Coding of Moving Pictures and Associated Audio Information: Video. ISO/IEC 13818-2:2000 (2000)
5. Standard NTSC Channels & Frequencies. http://radiotechnicalservices.com/tvchannels.pdf
6. Texas Instruments Inc., Spread Spectrum Clocking Using the CDCS502/503 Application Report, SCAA103 (2009). http://focus.ti.com/lit/an/scaa103/scaa103.pdf

Chapter 9
Memory Sub-system Design Considerations

The most critical bus in a DSP system today is the memory bus where a large amount of ultra-high-speed data is being transferred from the DSP to the physical memory devices and vice versa. The data on this bus are switching very fast. The rise and fall times of the data, memory clocks, and control signals are approaching sub-nanosecond range. These fast transients generate noise, radiation, power supply droops, signal integrity, and memory timing issues. This chapter covers memory sub-system design techniques to minimize the effects of high-speed data propagating.

9.1 DDR Memory Overview

It is assumed in this chapter that the memory is DDR memory, since DDR design presents many challenges as DDR transmits and receives data at both edges of the memory clock. These include a noise-sensitive analog circuit called Delay Locked Loop or DLL. In this case, the internal and external noise can cause excessive DLL jitter which leads to memory failures.

The three types of DDR memories shown in Table 9.1 are DDR1, DDR2, DDR3, DDR4, and LPDDR4 or Low Power DDR4. The transfer rate in DDR is generally defined as MT/s or Mega Transfers per Second. So DDR4-3200 is DDR4 data transfer rate of 3200 MT/s. In this case, the clock rate is 1600 MHz as each edge of the clock initiates one transfer.

Even though DDR has evolved from DDR1 to DDR4, soon to be DDR5, the fundamentals of DDR Write and Read Cycles remain essentially unchanged.

Figure 9.1 shows the basic DSP and DDR interface, and the signal definitions are in Table 9.2.

© The Author(s), under exclusive license to Springer Nature Switzerland AG 2023   115
T. T. Tran, *High-Speed System and Analog Input/Output Design*,
https://doi.org/10.1007/978-3-031-04954-5_9

Table 9.1 DDR SDRAM overview

Parameter	DDR1	DDR2	DDR3	DDR4	LPDDR4
Clock speed	Up to 200 MHz	Up to 400 MHz	Up to 933 MHz	Up to 1600 MHz	Up to 1600 MHz
Power consumption	High 2.5 V Core/IO	Moderate 1.8 V Core/IO	Moderate, less than DDR2, 1.5 V Core/IO	Moderate less than DDR3, 1.2 V Core/IO	Low 1.1 V Core, 1.8 V IO
Differential clock	Yes	Yes	Yes	Yes	Yes
Differential strobe, DQS	No	Yes (optional)	Yes	Yes	No
External V_{REF}	Yes, $0.49 \times V_{dd}$ Min, $0.51 \times V_{dd}$ Max	Yes, $0.49 \times V_{dd}$ Min, $0.51 \times V_{dd}$ Max	Yes, $0.49 \times V_{dd}$ Min, $0.51 \times V_{dd}$ Max	Yes, $0.49 \times V_{dd}$ Min, $0.51 \times V_{dd}$ Max	No, internal V_{REF}
DLL for DQ and DQS alignment	Yes	Yes	Yes	Yes	No, on-die terminations

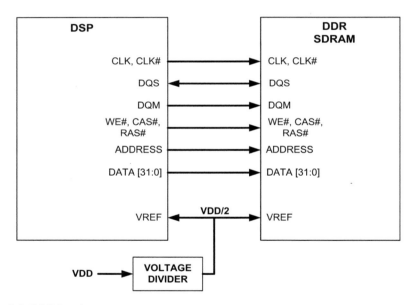

Fig. 9.1 DDR interface

9.1.1 DDR Write Cycle

Figure 9.2 shows a memory Write Cycle where $D0$ is the least significant data bit and DQS is the Data Strobe. The data transfers happen at both edges of DQS and DQS is typically a $90°$ phase-shift from the first data burst. Figure 9.3 shows an oscilloscope capture of the actual memory bus.

Table 9.2 DDR signal definition

CLK	DDR clock
CLK#	Inverted DDR clock
DQS	Data strobe: A bidirectional signal, output from DSP for write and input from memory for read.
DQM	Data mask: Input data is masked when DQM is high along with that input data during a write access.
WE#	Active low write enable
CAS#, RAS#	Active low column address and row address strobes
V_{REF}	Reference voltage: Half the supply voltage.
ADDRESS	Address bus: Provide the row address for ACTIVATE commands and the column address for READ/WRITE commands.
DATA [32:0]	32-bit Bidirectional data bus

Fig. 9.2 DDR write cycle

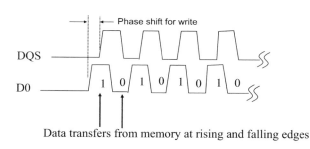

Data transfers from memory at rising and falling edges

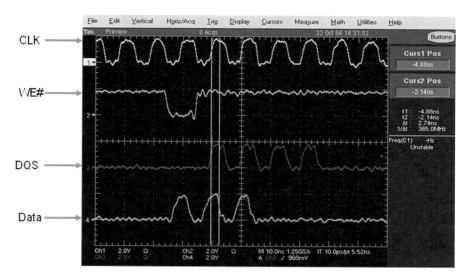

Fig. 9.3 DDR write cycle scope captured

Fig. 9.4 DDR read cycle

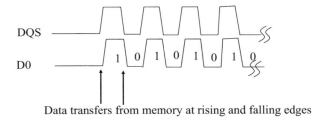

Data transfers from memory at rising and falling edges

9.1.2   DDR Read Cycle

Figure 9.4 shows a memory Read Cycle where D0 is the least significant data bit and DQS is the Data Strobe. The data transfers happen at both edges of DQS and DQS is synchronized with D0. In the Read Cycle, memory device outputs DQS and drives the data bus synchronously.

9.2   DDR Memory Signal Integrity

As covered in Chaps. 2 and 3, transmission line effects and crosstalk are results of bad signal integrity design. Since memory timing is so critical, the excessive over-shoots, undershoots, and glitches on the signal can false trigger and cause memory read and write failures. The worst part is that the memory controller such as DDR controller relies on a noise-sensitive analog circuit such as DLL to synchronize and delay the strobes to read the incoming data from external memory devices.

Simulated Figs. 9.5, 9.6, and 9.7 [1] demonstrate good, bad, and ugly signal integrity designs, respectively. The bad case in Fig. 9.6 has a glitch right at the switching threshold which may cause false clocking; the system may see this glitch as a high and low transition just like a clock input and respond to it.

For the ugly case in Fig. 9.7, the overshoots and undershoots are so excessive that the peak of the overshoot crosses over the minimum input high voltage, V_{ih}, and causes false clocking. Also, these overshoots and undershoots generate a lot of noise and radiation. The rule-of-thumb is to fine-tune the signal until all the overshoots are much lower than the threshold voltage, V_{ih}.

DDR2 Memory System Design Example 9.1
The following design rules for good memory signal integrity are:

- Apply good decoupling techniques as shown in Chap. 5. It is highly recommended to have one high-frequency capacitor (0.01–0.22 μF) per DDR power pin and one bulk capacitor (10 μF) for the DDR region. These decoupling techniques are also required for the memory devices in the design.

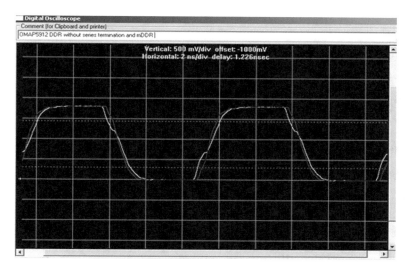

Fig. 9.5   Good DDR clock waveform

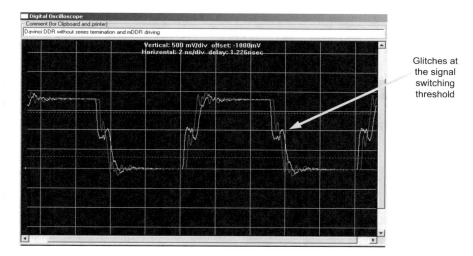

Fig. 9.6   Bad DDR clock waveform

- Add termination resistors on the data bus and the control signals. Where to place the resistors on the bidirectional bus depends on which device has higher drive strength. Place the resistors nearby the device with the higher drive strength. For example, DDR memory devices typically have strong buffers to allow for non-

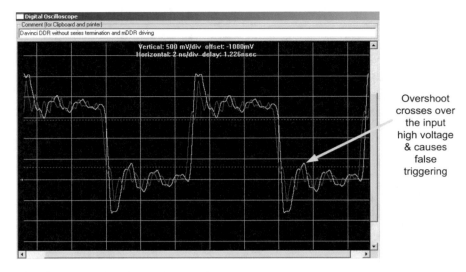

Overshoot crosses over the input high voltage & causes false triggering

Fig. 9.7 DDR clock with excessive overshoots

embedded designs such as PC. In this case, always put the termination resistors right by the memory devices. Ideally, add termination resistors at the output of the device driving the bidirectional bus. See the design examples in Figs. 9.8 [2] and 9.9.

- Isolate and decouple the DLL power supply and the V_{REF} voltage pins. For the DLL, follow the rules described in the PLL chapter, Chap. 6. To generate V_{REF}, use a resistor divider and divide the memory power supply voltage, Vdd, required for both DSP and memory devices. See the design example in Fig. 9.9 [3].

DDR4 Memory System Design Example 9.2   For DDR4-3200 or higher designs, it is difficult to manually characterize timings as done in Sect. 9.1, because specialized equipment such as Logic Analyzer is not readily available to measure pico-second timings. The industry now relies on automated design tools to analyze DDR memory buses and to verify timing margins. One example here is HyperLynx DDR Batch Simulation [1, 4], which runs worst-case analysis for data read/write and command/ address, and provides comprehensive reports with pass/fail tests, including eye diagrams as in Figs. 9.10 and 9.11.

In Fig. 9.10, the blue block in the middle of the eye diagram is the mask or the keep-out region in which any signal touching this mask indicates DDR timing violations. Figure 9.11 shows passes and fails of all the WRITE/READ cycles. For failed cases, engineers must find the root cause and fix the memory bus design to pass all WRITE/READ cycles before releasing the PCB to fabrication. The last step of memory validation is to run extensive memory tests on the actual hardware to verify timings and guarantee compatibility.

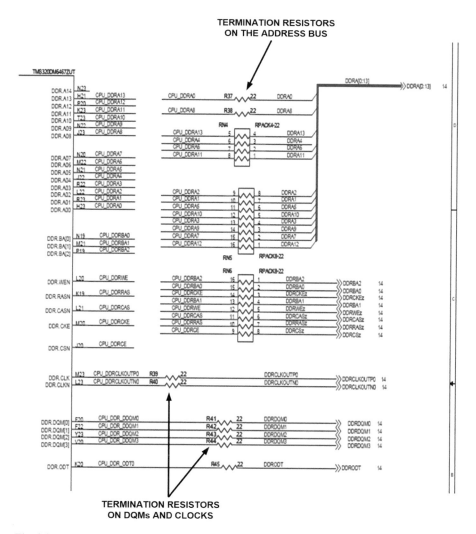

Fig. 9.8 DDR design example at the DSP

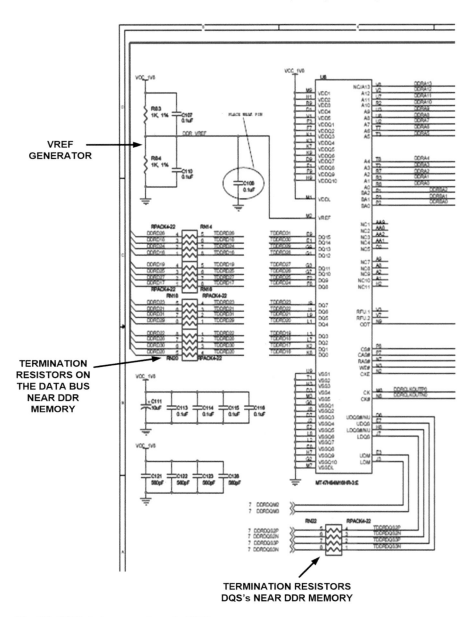

Fig. 9.9   DDR design example at the DDR memory

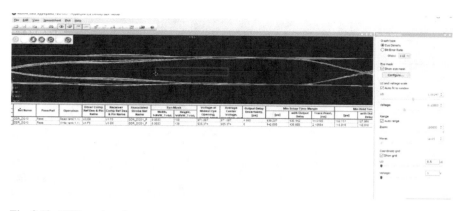

Fig. 9.10 DDR read cycle

Fig. 9.11 DDR write/ad pass and fail

References

1. Mentor Graphics, HyperLynx Signal Integrity Simulation Software (2004). http://www.mentor.com/products/pcb-system-design/circuit-simulation/hyperlynx-signal-integrity/
2. Texas Instruments Inc., TMS320DM6467 Digital Media System-on-Chip, SPRS403F (2007)
3. Micron, DDR2 SDRAM 1Gb: x4, x8, x16 DDR SDRAM (2009). http://download.micron.com/pdf/datasheets/dram/ddr2/1GbDDR2.pdf
4. Mentor Graphics, HyperLynx DDR Signal Simulations (2019). https://www.eeweb.com/taking-the-guesswork-out-of-ddr-design-with-integrated-schematic-layout-and-simulation-tools/

Chapter 10
USB 3.1 Channel Design

USB 3.1 is one of the latest industry standards which operates up to 10 Gbps speed. Designing USB channel requires extensive *s*-parameter simulations and running compliance tests. This chapter shows an example of how to use HyperLynx SERDES Compliance Wizard [1] tool to develop and simulate USB 3.1 channel, and to check USB compatibility.

10.1   USB 3.1 10 Gbps Channel Design

The design objective is to develop a USB channel that has a length of 18 in. between the USB HUB and USB Device and determine if USB Redriver is required to comply with USB standards. Figure 10.1 shows the block diagram of this design example.

To begin, use a PCB design software and design two differential pairs, and each pair has 100 Ω differential impedance. From the layout, extract the differential pairs and generate an *s*-parameter model, having 4 ports and 16 in. in length. Then, use LineSim in HyperLynx to simulate the channel. Since the two channels, TRANSMIT (TX) and RECEIVE (RX), are identical, only one channel is being designed to demonstrate the concept here. In practice, both channels must be modeled to make sure that the design is fully compliant.

The LineSim circuit in Fig. 10.2 is the USB channel design based on the datapath defined in the block diagram, Fig. 10.1.

As expected, the simulation results in Fig. 10.3 show the design failed Eye Height, Blue Mask (minimum open area required for USB) touching the border of the Eye Diagram. This concludes that a Redriver is required to compensate for the losses associated with a 16-in. channel.

Now, to add a Redriver to the design, a circuit in Fig. 10.4 was created in HyperLynx LineSim [1] consists of one TI TUSB1002A USB 3.2 Redriver [2], USB Transmit, USB Receive, and the PCB Channel. The Channel includes three

© The Author(s), under exclusive license to Springer Nature Switzerland AG 2023   125
T. T. Tran, *High-Speed System and Analog Input/Output Design*,
https://doi.org/10.1007/978-3-031-04954-5_10

Fig. 10.1 USB 3.1 block
diagram

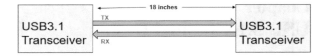

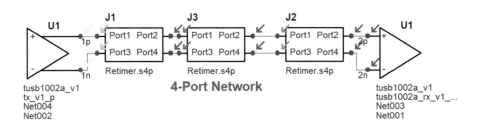

Fig. 10.2 HyperLynx LineSim circuit of USB 3.1 channel with no redrivers or retimers in the transmission paths. J1, J2 and J3 are s-parameter models of the channel extracted from the PCB layout.

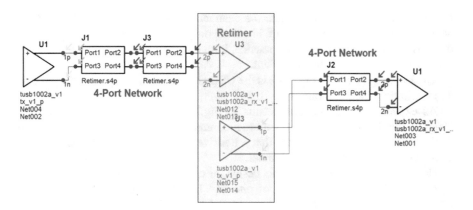

Fig. 10.3 USB 3.1 simulation results with no redrivers or retimers

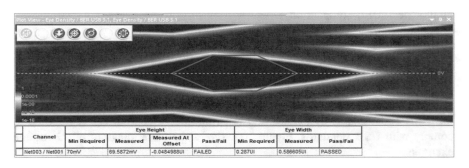

Fig. 10.4 USB circuit with redriver or retimer

s-parameter models, and each model is a four-port network 6 in. in length. The total channel is 18 in.

When the Redriver is inserted into the USB channel to compensate for the channel loss, the simulation must be done in a closed loop in which USB Transmit (U1) sends data to Redriver input (U3), and Redriver (U3) sends data output back to USB Receive (U1). In this configuration, one simulation reports the performance of the complete USB loop, transmit and receive paths.

The simulation results in Fig. 10.5 now show PASS (USB 3.1 compliance) on both paths, from U1 to U3 and U3 to U1. The Eye Width and Eye Height were measured at the Bit Error Rate (BER) of 1×10^{-12}.

Figure 10.6 shows the Eye Diagram of the USB channel from U1 to U3 that passes the compliance tests with plenty of margin. For example, the USB standard requires a minimum Eye Height of 70 mV while the design has an Eye Height of 464 mV. And the minimum Eye Width is $0.286 \times$ UI while the design has an Eye Width of $0.774 \times$ UI. UI is Unit Interval which is one divided by the bitrate. For USB 3.1 10 Gbps, the UI is 100 pS.

Fig. 10.5   USB with redriver simulation results

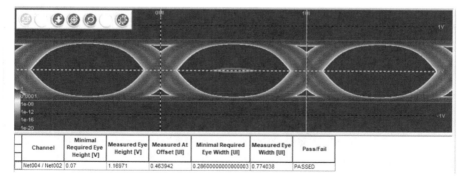

Fig. 10.6   Four-port *s*-parameter matrix

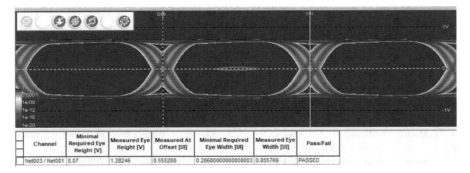

Channel	Minimal Required Eye Height [V]	Measured Eye Height [V]	Measured At Offset [UI]	Minimal Required Eye Width [UI]	Measured Eye Width [UI]	Pass/Fail
Net003 / Net001	0.07	1.28246	0.555288	0.2860000000000003	0.855769	PASSED

Fig. 10.7 Four-port mixed-mode *s*-parameters table

Again, Fig. 10.7 shows the Eye Diagram of the USB channel from U3 to U1 that passes the compliance tests with plenty of margin. Eye Width and Eye Height were measured $0.856 \times$ UI and $0.555 \times$ UI, respectively.

10.2   Layout Considerations of USB 3.1 10 Gbps Channels

PCB layout of high-speed channel is challenging as many components on the signal paths can greatly affect USB performance. These components include connectors, trace lengths, vias, and AC-coupling capacitors being in the circuit for isolating DC as described in Chap. 3. For general layout rules, refer to Chap. 14, and here are the most important USB layout rules.

- Skew within one differential pair, between Data+ and Data−, is 5 mils (5 thousandths of an inch) maximum length mismatch.
- Maximum number of vias on USB is 2 per trace, 2 for Data+ and 2 for Data−. One differential via must have one Anti-Pad, which is a keep-out area where no signal traces are allowed to be routed through. Also, each signal via must have a ground via next to it as shown in Fig. 10.8. The dimensions of the anti-pad and via depend on PCB stackup, PCB process requirements, and targeted via impedance. In general, the gap between the via diameter and the anti-pad is 5 or 6 mils.
- For high-speed signals, via stubs are not allowed and must be removed. The technique commonly being used to remove via stubs is back drilling. This is because via stubs cause signal reflections that lead to degrading insertion loss.
- For one differential pair, the trace width and the spacing between the positive data and the negative data need to be calculated to meet the USB differential impedance requirements. Many free impedance calculator tools are readily available, and one of the popular ones is https://www.allaboutcircuits.com/tools/filter/pcb/.
- For the spacing between two differential pairs, use a 5 W or five times the trace width rule [3].

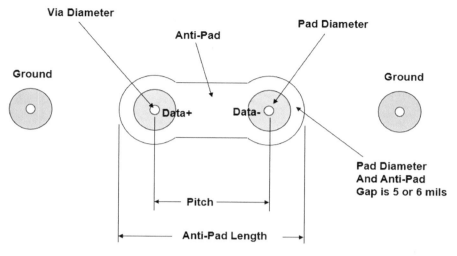

Fig. 10.8 Anti-pad for differential via

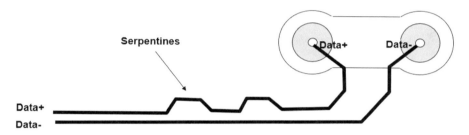

Fig. 10.9 Serpentines placed near the via with the shorter trace

- Always route the differential pair together and if bends are necessary, use rounded bends. Also, if serpentines need to be added to match the positive and negative signals, put the serpentines at the end near the area that causes the mismatch as shown in Fig. 10.9.

10.3   USB 3.1 10 Gbps Channel Design

As demonstrated in this chapter, designing high bitrate channels require carefully tuning and simulating the layout to pass the industry standard compliance tests. The good thing is that the tool technology has advanced to the point where many tools can be used to analyze designs with worst-case conditions and provide accurate results. As always, simulations are good but are not perfect, so building hardware and running tests are still necessary to guarantee 100% compatibility.

In this design example, only the channel on a PCB was analyzed. In practice, in addition to the channel, USB design must include all other losses associated with USB connectors, cables, transmitter, and receiver. The good thing here is that the simulation models of these components are readily available, so designers can incorporate the models in the simulations to validate if the complete design passes the compliance tests.

References

1. Mentor Graphics HyperLynx SERDES Compliance Wizards
2. Texas Instruments TUSB1002A USB3.2 10Gbps Dual-Channel Linear Redriver, Document Number: SLLSF63A, Mar 2018
3. Texas Instruments High-Speed Interface Layout Guidelines, SPRAAR7H, Aug 2014

Chapter 11
Phase-Locked Loop (PLL)

PLL is the heart of practically all electronic components and or modules where different clock frequencies are required to synchronize the data transmitting and receiving to and from externals, respectively. The input clock to the PLL is much lower than the DSP maximum clock frequency. PLL is typically used as a frequency synthesizer to generate the clock for the DSP core. For example, the input clock to the 1.2 GHz DSP [1] is 66 MHz.

PLL is an analog circuit that is very sensitive to power supply noise. Noise causes jitter and excessive jitter causes timing violations which lead to system failures. The two main PLL architectures are analog PLL (APLL) and digital PLL (DPLL). Understanding the differences helps to make the design tradeoffs often required for minimizing noise and jitter caused by external circuitries, such as the power supply and other noisy switching devices.

11.1 Analog PLL (APLL)

As stated, PLL generally functions as a frequency synthesizer, multiplying the input clock by an integer. This integer is a ratio of the feedback counter M divided by the input counter N as shown in Fig. 11.1.

Table 11.1 provides a brief description of each block shown Fig. 11.1 for the APLL.

The following provides an overview of how the PLL functions as a frequency synthesizer:

1. The reference clock is connected to the PDF input. The Divide-by-N counter reduces the input frequency.
2. The PDF compares the output of the Divide-by-M counter with the reference clock and generates an error signal.

© The Author(s), under exclusive license to Springer Nature Switzerland AG 2023
T. T. Tran, *High-Speed System and Analog Input/Output Design*,
https://doi.org/10.1007/978-3-031-04954-5_11

Fig. 11.1 Analog PLL

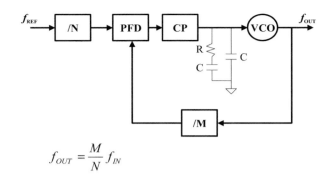

$$f_{OUT} = \frac{M}{N} f_{IN}$$

Table 11.1 Analog PLL description

Name	Description	Function
/N	Divide-by-N	Divide-by-N counter scales down the input frequency.
PFD	Phase-Frequency Detector	PFD compares the frequency and the phase of the input and the feedback clock signals and generates an error signal.
CP	Charge Pump	This is typically a constant current source controlled by the error signal output from the PFD block.
VCO	Voltage Controlled Oscillator	This VCO oscillates at a frequency controlled by the DC input voltage derived from integrating the error signal.
/M	Divide-by-M	Divide-by-M counter scales down the output frequency.

3. Based on the error signal, the CP charges or discharges the current store on the loop filter, an RC filter is shown in Fig. 11.1. This increases or decreases the VCO control voltage. For some PLL architectures, increasing the VCO control voltage increases the clock frequency and decreasing the voltage lowers the clock output frequency.

4. The phase correction continues until both the feedback signal from the Divide-by-M counter and the reference clock are synchronized. At this point, the error voltage should be zero.

5. The output clock frequency is equal to the ratio of the Divide-by-M counter and the Divide-by-N counter multiplied by the input clock frequency. As a rule-of-thumb, a higher multiplier ratio yields higher jitter, so keep the M and N ratio as low as possible when designing with PLLs. The PLL output frequency, f_{out}, for a given input frequency, f_{in}, is

$$f_{out} = \frac{M}{N} f_{in}, \qquad (11.1)$$

where M is the PLL feedback counter and N is the input counter.

11.2 PLL Jitter

Jitter in PLL design is defined as the signal timing displacement from a reference clock. The three main sources of DSP PLL jitter are jitter generated by the reference clock itself, power supply noise, and noise coupling from external and internal circuitries. The following lists important techniques for designers to minimize the DSP PLL jitter:

- Select a reference clock oscillator with the lowest jitter specification possible.
- Heavily filter the clock circuit to reduce the effect of noise on the output jitter. See the following section on PLL isolation.
- Use a series termination resistor at the output of the reference clock to control the edge rate.
- Distribute the clock differentially if possible. Differential signals reject common mode noise and crosstalk.
- Set the multiplier as low as possible to achieve maximum DSP operating frequency. Keep in mind that a higher multiply ratio yields higher output jitter.

In all cases, jitter can be minimized but cannot be eliminated. The three types of deterministic jitter [2] important for frequency synthesizers and DSP performance are long-term jitter, cycle-to-cycle jitter, and period jitter.

11.2.1 Long-Term Jitter

See Fig. 11.2 where long-term jitter is defined as a time displacement from the ideal reference clock input over a large number of transitions. Long-term jitter measures the deviation of a rising edge over a large number of cycles (N) after the first rising edge.

$$\text{Peak-to-Peak Jitter} = \text{Max Period}\,(N\ \text{cycles}) - \text{Min Period}\,(N\ \text{cycles}),$$

where Max Period is the maximum period equal to 1 divided by the operating frequency measured at N number of cycles and Min Period is the minimum period equal to 1 divided by the operating frequency measured at N number of cycles.

Fig. 11.2 Long-term jitter

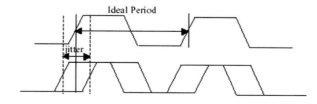

The long-term jitter can be measured using automatic jitter measurement equipment [3] or using a high-speed digital sampling scope. Here are the steps to measure long-term jitter using a scope:

- Use a high-speed 10 GHz sampling oscilloscope.
- Use the input clock to trigger the scope and set the scope in the Infinite Persistence mode.
- The deviation is measured from the first rising edge to the *N*th cycle. The "fuzz" shown on the scope in Fig. 11.3 is the long-term jitter.

11.2.2   *Cycle-to-Cycle Jitter*

See Fig. 11.4 where cycle-to-cycle is defined as the deviation of the clock period between two consecutive clock cycles.

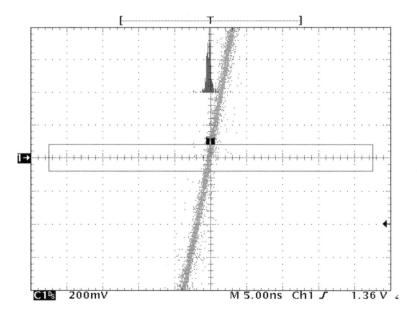

Fig. 11.3 Jitter measured by digital scope

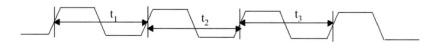

Fig. 11.4 Cycle-to-cycle jitter

In Fig. 11.4, the cycle-to-cycle is measured by subtracting t_2 from t_1, t_3 from t_2, and so on.

11.2.2.1 Cycle-to-Cycle Jitter Measurement

This is a difficult parameter to accurately measure with the high-speed sampling scope. Because the sampling scopes on the market today are not capable of measuring jitter in a few picoseconds range. The best way is to use a Timing Interval Analyzer (TIA) which captures one cycle at a time and compares the timing differences between two consecutive cycles. Another method is to use a scope with a cycle-to-cycle jitter measurement option. This method is outlined as follows:

- Use a high-speed 10 GHz sampling oscilloscope [4] with cycle-to-cycle jitter option.
- Trigger the PLL output clock and measure the cycle-to-cycle jitter. Use the windowing method to measure the changes from one cycle to another.

11.2.3 Period Jitter

See Fig. 11.5 where period jitter is defined as the maximum deviation in the clock's transition from its ideal position. These periods are non-successive.

11.2.3.1 Period Jitter Measurement

- Use a high-speed 10 GHz sampling oscilloscope [4].
- Set the scope in the Infinite Persistence mode and trigger the PLL clock output on the rising edge.
- Measure the "fuzz" shown on the screen at the next rising of the clock. Check indentation.

In summary, the jitter measurements can either be done using a high-speed digital sampling scope, Timing Interval Analyzer (TIA), or an automatic jitter measurement system [3].

Fig. 11.5 Period jitter

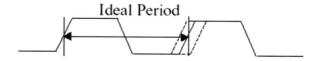

11.3   Digital PLL (DPLL)

The main differences between the APLL and DPLL are that the DPLL replaces the analog filter with a digital controller block that filters the phase error in the digital domain and replaces the VCO with a Digital Controller Oscillator (DCO). The advantages of the DPLL are:

- The DPLL supports a wide range of input frequencies from 30 kHz to 65 MHz or higher.
- The DPLL design requires a smaller silicon area to implement and consumes less power than the APLL.
- The DPLL does not have analog filter components such as capacitors which can cause leakage current. This leads to lower power consumption.
- The DPLL block is scalable and portable. The same design can be implemented on different process technology nodes.
- The DPLL design can be optimized for low jitter. But it may not be acceptable for jitter sensitive designs such as USB, audio, and video clocks.

The disadvantages of the DPLL are:

- It is very sensitive to external and internal power supply noise. Use of a linear regulator plus a Pi filter to isolate the power supply from the DPLL is recommended.
- Low power supply rejection ratio.
- In addition to power supply sensitivity, quantization noise and phase detector dead zone are the major sources of output jitter.
- Requiring a DAC block to control the oscillator. This makes the DPLL more sensitive to noise.

Figure 11.6 shows a typical DPLL architecture [5] and Table 11.2 describes the function of each block in the architecture.

Fig. 11.6   Digital PLL

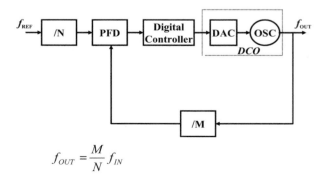

$$f_{OUT} = \frac{M}{N} f_{IN}$$

Table 11.2 Digital PLL description

Name	Description	Function
IN	Divide-by-*N*	Divide-by-*N* counter scales down the input frequency.
PFD	Phase-Frequency Detector	PFD compares the frequency and the phase of the input and the feedback clock signals and generates an error signal.
Digital Controller	Digital Controller	This digital filter block detects the phase error information and digitally controls the oscillator.
DCO	Digital Controlled Oscillator	This DCO converts the control code to analog levels and generates a stable clock output.
IM	Divide-by-*M*	Divide-by-*M* counter scales down the output frequency.

Table 11.3 APLL and DPLL jitter comparison

Noise on power supply	mV	100		100		100	
	Hz	100		10,000		1,000,000	
	Process	Pk-Pk (ps)	% jitter at max freq.	Pk-Pk (ps)	% jitter at max freq.	Pk-Pk (ps)	% jitter at max freq.
DPLL	Cold	201.21	2.90	219.12	3.16	199.38	2.87
Period jitter	Baseline	180.05	2.59	181.58	2.61	177.31	2.55
CVDD	Hot	156.45	2.25	149.33	2.15	160.32	2.31
APLL	Cold	195.11	1.87	192.06	1.84	195.11	1.87
Period jitter	Baseline	198.30	1.90	191.38	1.84	197.96	1.90
CVDD	Hot	178.02	1.71	182.29	1.75	173.75	1.67

11.4   APLL and DPLL Jitter Characterization

Table 11.3 shows a jitter comparison between an analog and a digital PLL that shows the effects of process variation where Hot is fast, Cold is slow and Baseline is typical. In this DSP design, the DPLL power supply is isolated by an internal low dropout regulator (LDO) while the APLL is connected directly to the common power supply plane. To test the noise sensitivity, 100 mV of noise modulating from 100 Hz to 1 MHz is injected into the power supply rails. The results showed that the peak-to-peak period jitter is less than 3% for the DPLL and is less than 2% for the APLL. With the LDO, the DPLL jitter is less than 4% up to 50 mV of noise on the power supply.

Designers need to be careful when injecting a signal onto the power supply to do jitter measurements. The nature of the signal used for simulating a noisy power supply condition can have a major impact on the PLL jitter. A squarewave signal with a frequency less than the PLL bandwidth characterizes the worst case PLL jitter. As far as the amplitude of the noise, the peak-to-peak voltage must be within the power supply limits. For example, for a 1.6 V $\pm$ 3% Core, the maximum acceptable peak-to-peak noise is 96 mV (-48 mV min and $+48$ mV max).

11.5   PLL Noise Isolation Techniques

As shown in previous sections, both an APLL and a DPLL are sensitive to noise, especially to noise frequency within the PLL bandwidth. PLL isolation is needed in order to prevent the high-frequency PLL signal from propagating out of the PLL section and affecting other circuits. PLL isolation can also attenuate the external noise propagating to the PLL circuit which causes excessive jitter. In many cases, external power supply noise causes the PLL to go unstable and the DSP to lock-up randomly.

11.5.1   Pi and T Filters

The two important filter schemes discussed in this document to isolate the PLL are low-frequency filtering and high-frequency filtering. For high-frequency filtering, a Pi or T network filter can be used as shown in Figs. 11.7 and 11.8:

The Pi filter circuit consists of one ferrite bead, L and two capacitors, C_1 and C_2. This circuit provides both input and output isolation where noise from the 3.3 V supply is attenuated by the ferrite bead and the C_2 capacitor and noise generated by the PLL circuit is isolated by the ferrite bead and the C_1 capacitor. Refer to Chap. 13 for the filter design and simulation information.

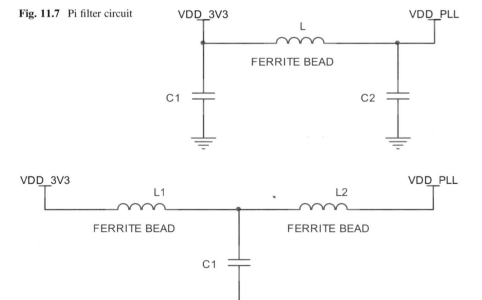

Fig. 11.7 Pi filter circuit

Fig. 11.8 T filter network

A T filter consists of two ferrite beads and one capacitor as shown in Fig. 11.8. Just like in a Pi filter, 3.3 V supply noise is attenuated by the L_1 ferrite bead and the C_1 capacitor and PLL noise is isolated by the L_2 ferrite bead and C_1 capacitor. Refer to Chap. 13 for the filter design and simulation information.

Both Pi and T circuits are good for filtering high-frequency noise but they are not as effective for low-frequency filtering since ferrite beads have almost zero AC impedance at low frequency. The Pi circuit has an advantage over the T circuit. Because this topology makes it possible to place the capacitor closer to the PLL voltage pin that ensures low impedance to ground and also the smallest current loop area, which reduces noise and EMI.

For low-frequency isolation, there are two common techniques, Pi filter with large bulk capacitor and linear voltage regulator.

One method for low-frequency filtering is shown in Fig. 11.9, where a resistor R replaces the ferrite bead and a bulk capacitor C_3 (10–33 μF) is added to the circuit. Low-frequency noise is attenuated by the resistor R and the bulk capacitor C_3. The resistor needs to be selected such that the voltage drop across the resistor is negligible. The low frequency −3 dB corner for this filter is approximated by Eq. (11.2). Notice that C_1 and C_2 are negligible in this case since its value is a lot lower than the bulk capacitor C_3.

$$f_{-3\mathrm{dB}} = \frac{1}{2\pi RC_3}. \tag{11.2}$$

Design Example 11.1   Design a PLL power supply filtering circuit that provides a 20 dB attenuation at 15 kHz. The tolerance for the PLL power supply is ±5% and the maximum current consumption is 10 mA.

Design steps are:

- The Pi filter circuit in Fig. 11.9 is a single pole filter neglecting C_1 and C_2. For a single pole filter, the attenuation is −20 dB/decade starting at the −3 dB corner frequency as shown in Eq. (11.2).
- $f_{-20\mathrm{dB}} = 10 \times f_{-3\mathrm{dB}}$, slope is 20 dB/dec so the frequency at −20 dB is equal to 10 times the frequency at −3 dB. Therefore, $f_{-3\mathrm{dB}} = (15 \text{ kHz})/10 = 1.5$ kHz.

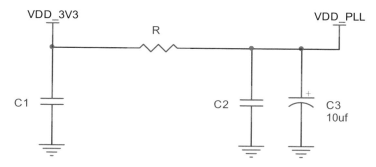

Fig. 11.9 Low-frequency Pi filter

- From Eq. (11.2),

$$f_{-3\mathrm{dB}} = \frac{1}{2\pi RC_3} = 1.5 \text{ kHz},$$

$$RC_3 = 1.06 \times 10^{-4},$$
$$\text{Let } R = 10 \text{ } \Omega,$$
$$C_3 = 10.6 \text{ or } 10 \text{ } \mu\text{F}.$$

- The voltage drop across the resistor is
  - $V = IR = 10 \times 10^{-3} = 0.01$ V. This is very small and is way within the power supply limits of 3.3 V $\pm$ 5%.

- The resistor power dissipation is
  - $P = VI$ where $V = IR \rightarrow P = I^2R = (10 \text{ mA})^2 \times 10 = 0.01$ mW. This small power dissipation allows designers to use a very small size resistor for this filter.

- Let C_1 and C_2 be a 0.01 μF capacitor since this is a good high-frequency decoupling capacitor as discussed in Chap. 13. The final circuit and simulation are shown in Figs. 11.10 and 11.11, respectively.

As shown in Fig. 11.11 simulation, the -3 dB corner frequency is at 1.5 kHz and the -20 dB attenuation is at 15 kHz. These are the design specifications.

In this design example, there is an IR voltage drop across the resistor R so it is very important to select the resistance to guarantee that the PLL supply voltage range is within the specified limits for worst case PLL current consumption.

11.5.2   Linear Voltage Regulators

Another method of low-frequency filtering is to use a linear voltage regulator. This method has the least effect on PLL performance. The linear regulator typically has good line regulation and power supply rejection characteristics which prevent

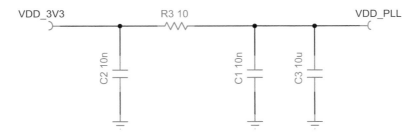

Fig. 11.10 Final Pi filter circuit for PLL

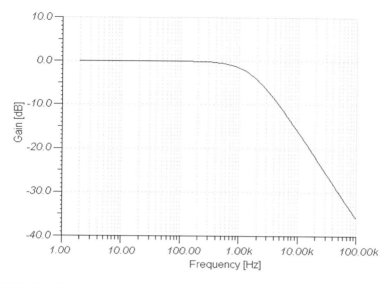

Fig. 11.11   Final Pi circuit simulation

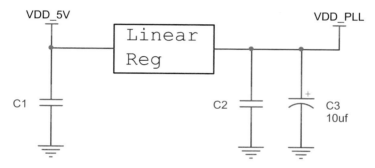

Fig. 11.12   Noise isolation with voltage regulator

low-frequency transients and high-frequency noise from entering the PLL circuit. The method shown in Fig. 11.12 is more expensive to implement than other methods described previously. But it is extremely effective in keeping the PLL voltage as clean as possible to guarantee lowest PLL jitter. Refer to Chap. 4 for design considerations.

One issue with using a linear regulator is that it does not reject high frequency very well. As shown in Fig. 11.13, the ripple rejection is approaching 0 dB (no rejection at all) for noise that is higher than 1 MHz. This high-frequency noise can cause more jitter in the PLL.

In summary, the best way to isolate PLL is using a combination of Pi filter and linear regulator. In this case, the Pi filter can be implemented with ferrite bead and

Fig. 11.13 Linear regulator
ripple rejection

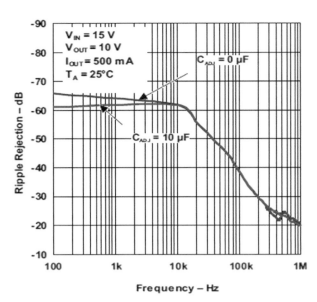

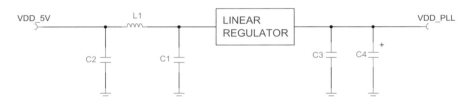

Fig. 11.14 Linear regulator Pi filter circuit

capacitors so there is no IR drop across the resistor as shown in the previous
example. The final circuit is shown in Fig. 11.14.

11.6 Summary

Because of low power consumption and fast response time, most of the PLL designs
integrated into the DSP today are based on digital PLL concepts. As discussed,
DPLL is very sensitive to power supply and input noise, so proper design noise
isolation filters are required to achieve the lowest jitter possible. The best approach is
using a combination of Pi filter and linear regulator as shown in Sect. 11.5.2. This
may not be possible due to PCB space limitations so designers must make design
compromises. If there is not enough room for the regulator circuit, then
implementing a Pi filter using a resistor instead of a ferrite bead is the second-best

approach. This has low-frequency and high-frequency filtering characteristics as demonstrated in Sect. 11.5.1.

References

1. Texas Instruments Inc., SM320C6455-EP Fixed-Point Digital Signal Processor. SPRS462B (2008), http://focus.ti.com/lit/ds/symlink/sm320c6455-ep.pdf
2. Cypress Semiconductor Corporation, Jitter in PLL-Based Systems: Causes, Effects and Solutions (1997)
3. Wavecrest, Examining Clock Signals and Measuring Jitter with the WAVECREST SIA-300. Application Note No. 142 (2002)
4. Agilent Technologies, Jitter Generation and Jitter Measurements with the Agilent 81134A Pulse Pattern Generator & 54855A Infiniium Oscilloscope (2003)
5. J. Lin, B. Haroun, T. Foo, J. Wang, B. Helmick, T. Mayhugh, C. Barr, J. Kirkpatrick, A PVT Tolerant 0.18MHz to 600MHz Self-Calibrated Digital PLL in 90nm CMOS Process, in *ISSCC* (2004)

Chapter 12
Power Supply Design Considerations

Power supply design is perhaps the most challenging aspect of the entire process of controlling noise and radiation in high-speed system design. This is largely because of the complexity of the dynamic load switching conditions. These include devices going into or out of low power modes, excessive in-rush current due to bus contention and charging decoupling capacitors, large voltage droop due to inadequate decoupling and layout, oscillations that overload the linear regulator output, and high current switching noise generated by switching voltage regulators. A clean and stable power supply design is required for all devices to guarantee system stability. This chapter outlines the importance of proper power supply design and the methods to minimize unwanted noise.

12.1 Power Supply Architectures

The two types of power supplies commonly being used in high-speed systems are linear and switching power supplies. The linear power supply has the best low noise characteristics typically required for analog audio, video, PLL, and data converter circuits. The disadvantages of this architecture are its power efficiency and dissipation. As shown in Fig. 12.1, the linear power supply consists of two main stages, input/output transistor and error amplifier. The input DC voltage here must be higher than the output voltage and the minimum input voltage varies depending on the component selected. So, it is important for designers to review the power supply's specifications and set input and output voltage levels appropriately.

The circuit in Fig. 12.1 operates as follows:

- The transistor T_1 operates in a linear region where the emitter current, I_e, (output current) is controlled by the base current, I_b, and the gain of the transistor, β.

© The Author(s), under exclusive license to Springer Nature Switzerland AG 2023   145
T. T. Tran, *High-Speed System and Analog Input/Output Design*,
https://doi.org/10.1007/978-3-031-04954-5_12

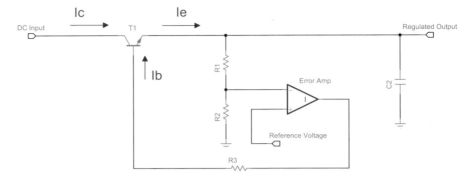

Fig. 12.1 Linear power supply architecture

$$I_e = I_c + I_b, \tag{12.1}$$

$$I_c = I_b\beta. \tag{12.2}$$

Substitute Eq. (12.2) into Eq. (12.1)

$$I_e = I_b\beta + I_b. \tag{12.3}$$

- If the output voltage drops due to higher current load, the error amplifier config-
  ured as a negative feedback circuit compares the Regulated Output divided by the
  resistors R_1 and R_2 to the Reference Voltage and drives higher base current I_b to
  maintain regulation. As shown in Eq. (12.3), the output current I_e is increased
  with the increase in the base current I_b.
- If the output voltage increases due to lighter current load, the error amplifier sees
  more negative voltage at the input and lowers the base current. This leads to lower
  output current and again the system maintains regulation.

Like any other feedback system, if there are changes in the input voltage and the
load current, the system takes some time to stabilize, and this time typically is
specified in the component datasheet under the transient response section. The
major issue with linear regulator is the power dissipation across the transistor T_1
for high output current applications. The power dissipation is

$$P_{T1} = (V_{in} - V_{out}) \times I_e. \tag{12.4}$$

For example, if the input voltage is 12 V and the output is regulated at 5 V as
shown in the Design Example 12.1. For the 1 A output current, the power dissipation
across the transistor T_1 is

$$P_{T1} = (12 - 5) \times 1 = 7 \text{ W}.$$

This power dissipation generates a lot of heat and increases the device operating
temperature to the point where heatsink is required to keep the device temperature

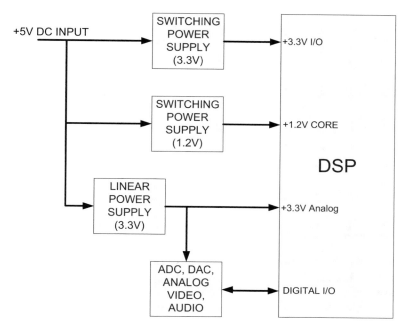

Fig. 12.2 DSP system power supply architecture

under the maximum allowable limits. As the current requirement increases with higher performance CPU/DSP and the system becomes smaller and smaller, it is no longer practical to use the linear regulators to generate all the supply voltages. In this case, it is best to use switching power supplies for the main power rails and linear regulators to provide clean low noise supplies to the noise-sensitive circuits such as analog and mixed analog/digital data converter circuits as shown in Fig. 12.2.

Design Example 12.1 Let us design a low noise high ripple rejection linear regulator to provide a +5 V to the audio circuit assuming +12 V is available on the board.
Design steps:

- Audio circuits are very sensitive to low-frequency noise, so it is best to select a regulator with a high power supply rejection ratio and with an external adjust pin to allow for additional decoupling. So, let us use LM317.
- As shown in the LM317 datasheet [1], this device has a Ripple Rejection specification of 62 dB minimum when using a 10 μF capacitor to decouple the Adjust pin.
- Figure 12.3 shows the complete LM317 circuit as recommended in the datasheet.
- In Fig. 12.3, C_1 prevents high-frequency noise from affecting the LM317 performance. D_1 and D_2 are diodes needed to discharge the currents in C_2 and C_3 to avoid these currents discharging into the output of LM317 during powering up and down of the regulator. These diodes are reverse based on normal operation. C_2 is required to get better Ripple Rejection specification and C_3 is a typical

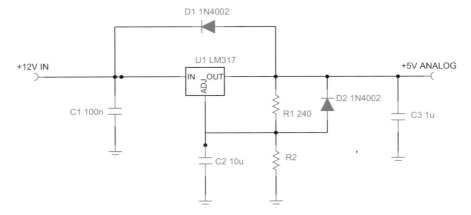

Fig. 12.3 Linear regulator circuit

decoupling capacitor. Keep in mind that C_3 does not need to be a large decoupling capacitor because LM317 is already doing a good job rejecting low-frequency noise.

- R_1 is fixed at 240 Ω and R_2 is to be calculated as follows. From the LM317 datasheet,

$$V_{out} = V_{ref}\left(1 + \frac{R_2}{R_1}\right) + \left(I_{adj}R_2\right),\qquad(12.5)$$

where I_{adj} is 50 μA and V_{ref} is 1.25 V. V_{out} is 5 V and R_1 is 240 Ω.

$$5 = 1.25\left(1 + \frac{R_2}{240}\right) + \left(50 \times 10^{-6} \times R_2\right).$$

For this application, it is acceptable to neglect 50 μA and solve for R_2. Therefore,

$$5 = 1.25\left(1 + \frac{R_2}{240}\right),$$

$R_2 = 720$ Ω. Put the complete circuit shown in Fig. 12.3 in the circuit simulator [2] and the results are shown in Fig. 12.4 where the output is regulated at +5 V when the input is +12 V.

As indicated earlier, for high current consumption circuits, it is best to go with switching power supply architecture. Because it provides much better power efficiency and lower power dissipation as compared to linear power supply. However, this architecture generates excessive output switching noise that can degrade system performance and cause EMI failures if designers are not carefully controlling the switching noise by applying proper design, PCB layout, and isolation techniques.

The two types of switching power supplies are "buck" and "boost" converters. The buck converter requires the input voltage to be higher than the output voltage; it

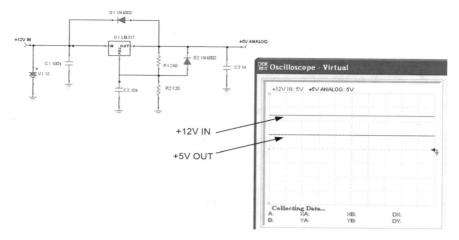

Fig. 12.4 LM317 circuit simulation results

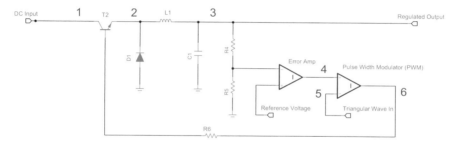

Fig. 12.5 Buck converter architecture [3]

is also known as a step-down converter. And the Boost converter generates an output that is higher than the input voltage; this is also known as a step-up converter. For high-speed systems, as many different voltages are required and these voltages can easily and economically be derived from the highest input voltage, a buck converter is preferred. In a buck converter as shown in Fig. 12.5, the three main stages are: (1) the power transistor stage; (2) the error amplifier stage; and (3) the pulse width modulator or PWM stage. The converter operates as follows:

- The power transistor T_2 operates in a saturation region where the transistor is being driven fully on or off. When the transistor is on, there is only a small resistance, Rdson in milli-ohms range, in series with the input and the output LC filter. This leads to very low power dissipation and high efficiency.
- The emitter output of the T_2 transistor is a digital waveform with a variable duty cycle controlled by the PWM circuit. This emitter output is filtered by the L_1 and C_1 and the output of the filter is a regulated DC output with some switching noise modulated on it.

- The regulated DC output is fed back to the error amplifier circuit through the resistor divider R_4 and R_5. This is a negative feedback loop so if the DC voltage increases, the error amplifier output will be driven more negative. If the DC voltage decreases, the error amplifier output will go more positive.
- The output of the error amplifier is an input to the PWM stage. This PWM compares the error amp voltage to a sawtooth waveform. If the error amp voltage increases, lower regulated DC output voltage, the PWM generates a higher duty cycle waveform to drive the power transistor. This leads to an increase in the regulated output voltage to maintain regulation. If the error amp voltage decreases, higher regulated DC output voltage, the PWM generates a lower duty cycle waveform to drive the power transistor. This leads to a decrease in the regulated output voltage to maintain regulation.
- The regulated output voltage is $V_{out} = V_{dc}\left(\frac{T_{on}}{T}\right)$, where T_{on} is the high time and T is the period.

Figure 12.6 demonstrates how the system compensates for higher output voltage. Higher output voltage leads to lower duty cycle signal and therefore lowers the output voltage back to the regulated level.

Figure 12.7 shows that the duty cycle of the signal increases when there is a decrease in the output voltage. This again forces the system back into regulation.

Design Example 12.2 Let us design a buck converter power supply for a DSP system that has the design specifications shown in Table 12.1 [4].

Refer to [4] to calculate all the component values shown in Fig. 12.8.

Fig. 12.6 Timing diagrams for higher error voltage

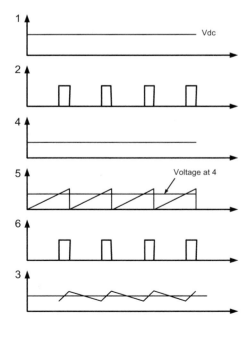

Fig. 12.7 Timing diagrams for lower error voltage

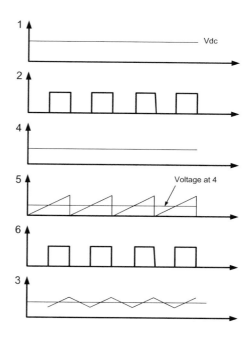

Table 12.1 Buck converter design specifications need to reformat

Parameter	Test conditions	Min	Typ	Max	Unit
Input					
V_{in} (input voltage)		10.8	12.0	13.2	V
I_{in} (input current)	$V_{in} = 12$ V, $I_{out} = 10$ A		1.7	1.8	A
UVLO_OFF (V_{in} undervoltage lockout turn off)	0 A $\leq V_{out} \leq 10$ A	5.4	6.0	6.6	V
UVLO_ON (V_{in} undervoltage lockout turn on)	0 A $\leq V_{out} \leq 10$ A	6.6	7.0	7.6	V
Output					
V_{out} (input voltage range)	$V_{in} = 12$ V, $I_{out} = 5$ A	3.1	3.3	3.5	V
Line regulation	$10.8 \leq V_{in} \leq 13.2$ V			0.5	%
Load regulation	0 A $\leq I_{out} \leq 10$ A			0.5	%
V_{ripple} (output ripple)	$V_{in} = 12$ V, $I_{out} = 10$ A			100	mVpp
I_{out} (output current)	$10.8 \leq V_{in} \leq 13.2$ V	0	5.0	10	A
I_{ocp} (output over-current inception point)	$V_{in} = 12$ V, $V_{out} = (V_{out} - 5)$	14	20	43	A
Transient response load step	10 A $\leq I_{out}(max) \leq 0.2 \times (I_{out}(max))$	8			A
Switching frequency		240	300	360	kHz
Peak efficiency	$V_{in} = 12$ V, $0 \leq I_{out} \leq 10$ A	90%			
Efficiency at full load	$V_{in} = 12$ V, $I_{out} = 10$ A	87%			
Operating temperature	$10.8 \leq V_{in} \leq 13.2$ V, $0 \leq I_{out} \leq 10$ A	−40	25	85	°C

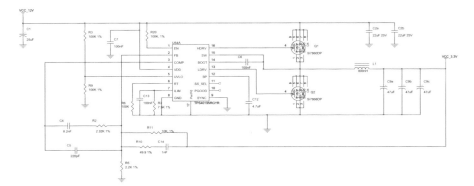

Fig. 12.8 3.3 V output buck converter schematic

The performance of this buck converter depends on the component values, the component placements, and the layout. Here are some important points to remember.

- Always follow the manufacturer design guidelines and layout.
- Place the switching power supply circuit at a corner of the PCB away from the rest of the system components.
- Keep all the switching current loops as small as possible. Refer to the manufacturer datasheet to figure out the possible current loops.
- In general, switching power supplies like Buck converter have y high current switching characteristics and this generates harmonics as high as 10–100 times the fundamental frequency. Proper shielding, decoupling, and isolating methods outlined in the manufacturer datasheet and in this book must be taken into consideration in order to increase the probability of success.

12.2   System Power Supply Architectural Considerations

Designing power supply is not just designing the power supply itself but is necessary to implement a system power to guarantee minimum noise and radiation to achieve higher performance, higher reliability, and lower cost. Generally, systems with low noise yield lower cost, because noisy systems tend to fail at a higher rate in manufacturing. In many cases, system designers had to take an expensive approach to solve issues by unnecessarily redesigning the mechanical chassis to add better shielding to reduce noise. This type of costly activity could have been prevented by applying good low noise design techniques right at the beginning.

It is understandable that redesigning electrical systems for low noise and EMI is difficult for engineers who are not familiar with the latest high-speed design techniques outlined in this book.

Assuming that the power supply itself was done properly as shown in the previous sections, the power integrity depends on how far the CPU/DSP is placed

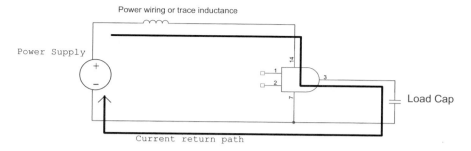

Fig. 12.9 Power supply current return path example 1

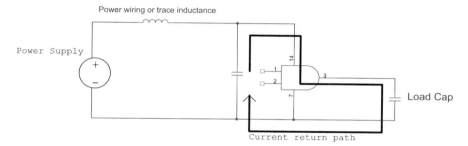

Fig. 12.10 Power supply current return path example 2

away from the power supply module and how well the CPU/DSP is being decoupled. Figures 12.9 and 12.10 show two circuits, where Fig. 12.10, has a decoupling capacitor close to the device. Assuming the same power supply trace inductances, Fig. 12.9 has a larger dynamic current return path leading to a larger power supply voltage droop and greater electromagnetic radiation. This may cause random system failures that are difficult to resolve. Refer to Chap. 13 for details of how to implement proper power supply decoupling capacitors to suppress noise while improving system reliability and performance.

One of the most challenging tasks for system designers is determining an acceptable noise level in a system. CPU/DSP data manuals clearly specify operating conditions but cannot account for the dynamic nature of high-speed systems. This is because the dynamic switching characteristics depend on how fast the system is turning on or off and how the loads are changing during transient conditions. The following are some of the important issues that must be addressed during the power supply design process:

- Power supply transient response, such as load regulation, line regulation, power supply ripple, power supply noise rejection, and power sequencing for multiple rails.
- Power supply decoupling to ensure minimum voltage droop at the pins of the CPU/DSP.

- Linear regulator versus switching regulator.
- Power supply planes versus power supply traces.
- CPU/DSP in-rush currents during power supply ramp and at steady-state.
- Power cycling: no residual voltage during CPU/DSP startup.
- Power supply rails sequencing: Core before IO or IO before Core.
- Be cautious with using a switching regulator to power the PLLs, audio CODECs, and video encoders and decoders.
- Always asserting reset during power supply ramp to reduce the probability of internal CPU/DSP bus contention.

Excessive power supply noise can have the following harmful effects:

- Voltage droop, inadequate decoupling capacitors, or current starvation may cause random logic failures. This is difficult to debug and may even require a redesign of the system to get rid of the noise.
- Inadequate voltage regulation can cause reliability problems or unintentional system shutdown.
- Excessive jitter may appear on clock circuits, especially the PLL.
- Radiation may rise to a level that makes it difficult to pass EMC tests.
- Visible and audible artifacts on video and audio systems.

Designers have three primary methods to overcome these problems: voltage regulator design (linear versus switcher), decoupling techniques, and PCB layout. One of the most important decisions made by designers is whether to use linear regulators or switching regulators. This decision requires a good understanding of the characteristics of the power supply and the impact of the supply on the system noise performance. The design of the power supply itself was covered in the previous sections. Let us look at the differences between linear and the switch regulators:

Table 12.2 helps determine which power supply solution is a better fit for the application. The next step is to determine the current consumptions and whether

Table 12.2 LDO versus switching regulators

Linear or low drop out (LDO) regulators	Switching regulators
Low noise with high power supply rejection ratio	Switching noise may cause EMI problems or video and audio artifacts
Fast response to load changes, typical 1 μs	Slow response to load changes
Low efficiency, typically 56%, may increase power dissipation, heatsinks may be required	High efficiency, typically 92%, provides low power dissipation
Unstable if the total decoupling capacitance is higher than the maximum allowable limit	DSP decoupling capacitor has a little or no impact on supply stability. On the other hand, PCB layout is critical
Excellent choice for video, audio, analog, and PLL circuits	Excellent choice for the core CPU and the IO power
Low cost	Higher cost due to the need for external filter components such as an LC filter at the output of the switch

power sequencing is required. In general, CPUs/DSPs have a minimum of two power supply rails, Core and IO. The sequence of ramping the Core and IO voltages can affect the startup current consumption so refer to the device data manuals to help design a robust power supply for a particular DSP. Here are recommended rules for selecting/designing a power supply.

12.2.1   CORE Voltage Regulator Design

- Refer to the device data manual to get the maximum current consumption for the Core supply. Many of the devices come with a power spreadsheet that can be used to estimate the current consumption of a particular CPU/DSP operating condition.
- Select a regulator with at least two times the maximum Core current capability. This provides an adequate margin to handle the dynamic current conditions.
- Be cautious with the current starvation condition. During startup, the surge current may exceed the maximum limit of the regulator for a short period of time. The selected regulator should have a soft-start capability to prevent thermal or over-current shutdown conditions from occurring.
- The final design step for the Core voltage regulator is whether a heatsink is required.

12.2.2   IO Voltage Regulator Design

IO voltage regulator design depends on the external loads in the specific application. For fast switching signals, the IO currents are supplied by the decoupling capacitors, not by the regulator itself due to the parasitic inductance associated with the power supply trace or plane. The dynamic current calculation will be shown in the decoupling section. The following guidelines provide a conservative method to design an IO voltage regulator for the DSP itself. It should be noted that this method applies to the DSP power alone as opposed to the entire system.

- Count the number of outputs from the CPU/DSP. All GPIOs should be considered as outputs.
- Multiply the number of outputs by the source current specified in the data manual.
- Add the total source current using the maximum IO current consumption specified in the data manual.
- Then, multiply the result by 2 to provide a 100% margin.
- Due to transmission line effects, IO current may surge during switching but this condition will be absorbed by the local CPU/DSP decoupling capacitors.
- The final step is to determine if heatsink is required or not.

Once designers complete the power supply architecture for a particular CPU/DSP, the next step is to determine if the CPU/DSP requires sequencing. The supply rails may be sequenced, for example, to ramp up the Core before the IO or vice versa. Proper sequencing is necessary to avoid internal contention.

- Improper reset during power-up. Reset must be asserted longer than the minimum reset pulse specified in the data manual.
- Core and IO not coming up within the specified time limits. Typically, DSPs do not require a power sequence but there is a time limit for one supply rail to be on while the other is off.
- Improper reset of the JTAG emulation port. For example, TRST needs to be stable and low. Excessive noise coupled with this signal may cause a startup problem or bus contention.
- Boot mode configuration pins are not being driven to proper states before releasing reset. Refer to the device data manual to make sure that the configuration pins have proper pull-ups and pull-downs and these pins have reached a stable logic level before releasing reset.

Figures 12.11 and 12.12 show two CPU/DSP power supply architectures [5], in Fig. 12.11, the Core and IO rails are powered up synchronously and in Fig. 12.12 the Core supply rail is ramped up before the IO. Refer to [6] to obtain more details about power management architectures.

12.3   Power Sequencing Considerations

The number of power supply rails required for CPU/DSP is increasing constantly as more and more peripherals are being integrated. And managing these rails during powering up or down of the DSP is a very difficult task. Typically, a high-performance CPU/DSP consists of at least three power supply rails, +1.8 V for

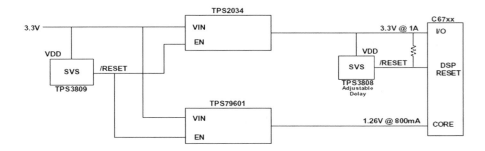

Fig. 12.11 Synchronous core and IO supply rails example

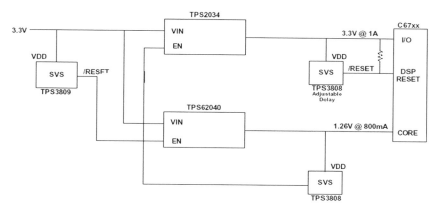

Fig. 12.12   Before IO example

DDR2, +3.3 V for data converters, +1.2 V for core. The internal logic has many voltage translations to enable all the blocks communicating to each other. During power-up, if one power supply rail goes up before another for some period, the internal logic can get to an unknown state which can cause internal bus contention and the system to go unstable. Designers must refer to the CPU/DSP datasheet and design in the power sequence if it is required. The problem is that the power supply sequence circuits shown in Figs. 12.11 and 12.12 can only guarantee the power-up sequence of the power supply itself, not the whole system. This is because the decoupling capacitors used around the CPU/DSP as shown in Fig. 12.10 affect the time it takes for the power supply to ramp up to the final operating voltage. This time can be calculated as follows:

$$I_{\text{power}} = C_{\text{decoupling}}\left(\frac{dV}{dt}\right),$$

$$dt = C_{\text{decoupling}}\left(\frac{dV}{I_{\text{power}}}\right), \tag{12.6}$$

where $C_{\text{decoupling}}$ is the total decoupling capacitance, I_{power} is the current sourced from the power supply, dV is the change in voltage, and dt is the time it takes to get to the dV value.

As shown in Eq. (12.6), for a given power supply, the time it takes to reach the final voltage level depends on the total decoupling capacitance. So, to guarantee a particular sequence, it is very important for designers to do the following:

- Refer to the DSP datasheet and determine whether power sequencing is necessary.
- If the selected DSP requires a power-up sequence, then use a topology like the one shown in Fig. 12.12 to develop the power supply system.

- Use Eq. (12.6) and calculate the ramp time for each power supply rail and verify that the power-up sequence was achieved at a system level with all the decoupling capacitors installed on the board.
- Use a current probe and measure the power-up currents (Core and IO) to make sure that there are no bus contentions. Keep in mind that all capacitors appear like a short circuit to ground when they start from a zero-volt state, so the surge current may be higher than expected during startup. Be sure to provide adequate margin in the power supply design to avoid false-triggering the over-current protection circuits. A good rule-of-thumb is adding a 50% margin to the maximum current consumed in the design.
- If there was excessive current consumption during power-up, designers need to check the following.
  - Is the system reset active? Reset signal should be asserted during this time.
  - Is the power-up sequence, correct?
  - Are any of the inputs left floating? All the inputs need to be pulled-up or down, but they cannot be floating. Some of the DSPs have internal pull-ups or downs integrated but not all of them, so make sure to check the data sheets and enable the pull-ups and downs appropriately.
  - Is the clock output of the DSP started running immediately after the power supply reaches its operating range? If not, check the emulation reset signal to make sure that this input is being driven correctly. If not, noise can couple to this input and randomly put the device in test modes.

Figure 12.13 summarizes the steps to verify the power sequencing and to debug the system startup over-current conditions.

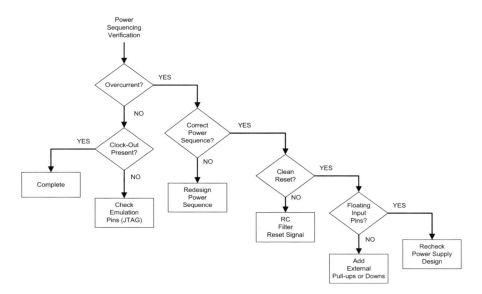

Fig. 12.13 Power sequencing verification

12.4 Summary

As demonstrated in this chapter, selecting the right power supply architectures for the high-speed system including surrounding analog/digital circuits and doing the system floorplan design are the two most important critical tasks for designers to get done first before getting started on the actual implementation. Good power integrity is key for achieving a low noise and low EMI system design and here is a list of recommendations to improve the probability of success:

- Develop a detailed system block diagram showing all the power supply requirements for all the components (CPU, DSP, ADC, DAC, video, audio, PLLs, DDR, etc.)
- Apply the techniques described in this chapter and calculate the current requirements for all the blocks. It is recommended to add a 50% margin to the overall current budget as this helps the system to better-handle dynamic situations.
- Highlight the noise sensitive circuits such as ADC, DAC, analog video/audio, and PLL and isolate these circuits by using high power supply rejection linear regulators if possible. Avoid powering these circuits with switching regulators.
- Do the floorplan design. Place the switching power supplies far away from analog and high-speed circuits. The best place for noisy power supplies is at the corner of the PCB.
- Select the power supply topologies and begin circuit implementation and layout. Refer to Sect. 12.2.1 for power sequencing.

References

1. Texas Instruments Inc., LM317 3-Terminal Adjustable Regulator (2008), http://focus.ti.com/lit/ds/symlink/lm317.pdf
2. Texas Instruments Inc., Spice-Based Analog Simulation Program (2008), http://focus.ti.com/docs/toolsw/folders/print/tina-ti.html
3. P. Abraham, *Switching Power Supply Design* (McGraw-Hill, New York, 1991)
4. Texas Instruments Inc., TPS40195 4.5-V to 20-V Synchronous Buck Controller with Synchronization and Power Good (2008), http://focus.ti.com/lit/ds/symlink/tps40195.pdf
5. Texas Instruments Inc., SM320C6713-EP Floating Point Digital Signal Processors (2009), http://focus.ti.com/lit/ds/symlink/sm320c6713b-ep.pdf
6. Texas Instruments Inc., Power Management Guide (2008), http://focus.ti.com/lit/sg/slvt145h/slvt145h.pdf

Chapter 13
Power Integrity

Poor power integrity is one of the most common root causes of system-related problems. This is because there are too many things that could affect power delivery to multiple devices on a system. These include DC resistance of PCB traces or power/ground planes, AC impedance of PCB traces, power supply decoupling around the DSP, and or other surrounding circuits such as DDR, clocks, and analog-to-digital and digital-to-analog converters. One of the most challenging tasks for designers is to determine the best decoupling techniques to achieve low noise and high performance. In general, component manufacturers provide a conservative recommendation for power supply decoupling, but in many cases, it is not practical to follow this recommendation because of PCB space availability, power consumption, EMI, or safety requirements. Also, component manufacturers always provide development platforms for designers to evaluate and these platforms typically are a lot larger than the actual design and are not required to be FCC certified, so copying what was done on the development platform is not a guarantee that the design will be successful. This chapter will discuss five important topics for designers: (1) DC resistance of traces; (2) AC impedance; (3) a general rule-of-thumb decoupling method; (4) an analytic decoupling method; and (5) how to make design tradeoffs to achieve the best noise performance possible.

13.1 Power Supply Decoupling Techniques

Once designers select and design a power supply for the DSP, the next step is to determine the decoupling capacitors needed to ensure that the power supply droop under all dynamic operating conditions is lower than the specified limits. For example, a 5% tolerance rating on a 3.3 V IO supply requires the ripple to be less than 165 mV. Let us first consider the situation where no decoupling capacitor is used as shown in Fig. 13.1.

© The Author(s), under exclusive license to Springer Nature Switzerland AG 2023
T. T. Tran, *High-Speed System and Analog Input/Output Design*,
https://doi.org/10.1007/978-3-031-04954-5_13

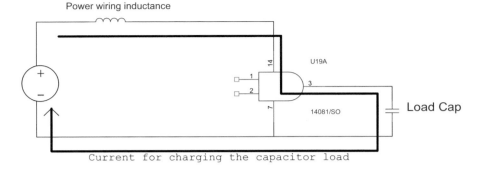

Fig. 13.1 DSP power w/o decoupling capacitor

In Fig. 13.1, the DSP labeled U19A is driving a capacitive load and is switching at a fast rate. Now, let us assume that the regulator is placed 5 in. away from the DSP and is routed with a 5-mil trace to the DSP. During fast switching, the power supply trace becomes an open circuit because of the parasitic inductance associated with the trace. This generates a large voltage droop at the pin of the DSP which can be estimated as follows [1].

$$\text{Droop} = L\left(\text{Max}\frac{dI}{dt}\right), \tag{13.1}$$

where L is the parasitic inductance.

$$\text{Max}\frac{dI}{dt} = \frac{1.52\Delta V}{(T_r)^2}C, \tag{13.2}$$

where ΔV is a switching voltage, C is a load capacitor, and T_r is a rise time.

For a 5-in. trace, the inductance is estimated by the signal integrity simulator [2] to be 600 nH/m. The inductance L is

$$L = 5 \text{ in.} \times 2.5 \text{ cm/in.} \times 1 \text{ m/100 cm} \times 600 \text{ nH/m} = 75 \text{ nH.}$$

Let $T_r = 2$ nS, $C = 50$ pF (load capacitance) and $\Delta V = 80\%$ of 3.3 or 2.64 V. The maximum calculated droop is

$$\text{Droop} = L\left(\text{Max}\frac{dI}{dt}\right) = L\left[\frac{1.52\Delta V}{(T_r)^2}C\right] = 75 \times 10^{-9}\left[\frac{1.52(2.64)}{(2 \times 10^{-9})^2} \times 50 \times 10^{-12}\right]$$

$$= 3.76 \text{ V.}$$

This example demonstrates that for a 5-in. trace, 2 ns signal, 50 pF load, and 3.3 V IO, the maximum power supply droop is 3.76 V. This level of droop is certain to cause random system failures. To compensate, decoupling capacitors are placed close to the DSP to provide the required charge during switching. What is the best method to filter the noise from the DSP system? Noise characteristics differ so much from system to system that no one method guarantees low noise and low radiation for all cases. However, designers can apply best practices outlined here to minimize the noise and to improve the probability of success. Before going into the decoupling techniques, it is important to understand the characteristics of the common components (capacitors, inductors, and ferrite beads) being used to filter out the power supply noise.

13.1.1   Capacitor Characteristics

The key specification of a capacitor used for decoupling is the self-resonant frequency. The capacitor remains capacitive below and starts to appear as an inductor above this frequency. Here is a series equivalent circuit representing the capacitor (Fig. 13.2).

The series equivalent circuit for a capacitor has three different components: equivalent series resistance (ESR), equivalent series inductance (ESL), and the capacitance itself. The self-resonant frequency happens at the point where the impedance of the capacitor, C, is equal to the impedance of the inductor, L.

$$Z_C, \text{capacitor} = \frac{1}{\omega C},$$

where C is capacitance and ω is 2π times the frequency, f.

Fig. 13.2 Capacitor
equivalent circuit

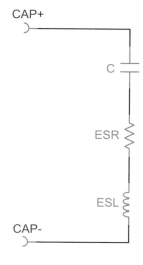

$$Z_L, \text{inductor} = \omega L,$$

where L is inductance.

At resonance, Z_L is equal to Z_C or

$$\frac{1}{\omega C} = \omega L,$$

$$\omega^2 = \frac{1}{LC},$$

$$\omega = \frac{1}{\sqrt{LC}},$$

where $\omega = 2\pi f$.

Therefore, the self-resonant frequency is

$$f_R = \frac{1}{2\pi\sqrt{LC}}. \qquad (13.3)$$

As shown in the self-resonance equation, lower capacitance and lower inductance yield a higher resonant frequency. For a given capacitance value, choosing a smaller surface mount component achieves a higher self-resonant frequency. Because a smaller component package typically has lower parasitic and lead inductance. The whole decoupling concept is to provide a low impedance path from the power supply to ground and to shunt the unwanted RF energy. This means that choosing a capacitor with high capacitance but with low inductance is important. The problem with this concept is that higher capacitance comes in a larger package which yields higher parasitic inductance. In many cases, it is better to use many capacitors with different values to decouple the DSP.

Figure 13.3 shows the capacitor frequency response. For a particular capacitor, the impedance decreases with frequency and reaches the lowest impedance point at resonant frequency, f_R. For frequencies above the resonant frequency, the impedance of the capacitor is dominated by the parasitic inductor, ESL. This causes the impedance to increase with frequency. It is recommended to operate in the capacitive region of the curve as this region guarantees a close to ideal impedance response of the capacitor.

There are many different types of capacitors and which type to use depends on the voltage, temperature, and frequency of the design. For example, low-frequency filtering requires a large electrolytic aluminum or tantalum capacitor with a value of 10 µF or higher, while high-frequency filtering needs a small film or ceramic capacitor with a value of less than 10 µF. Selecting the wrong capacitor type can negatively affect the performance of the system, so designers must carefully review the component specifications and the applications before making the selection. Table 13.1 shows the electrical characteristics of different types of capacitors commonly being used.

Fig. 13.3 Capacitor impedance response

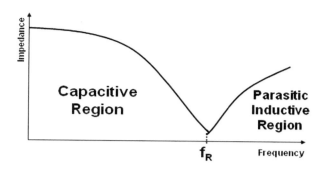

Table 13.1 Capacitor characteristics [6]

Capacitor type	Capacitance range	Characteristics	Typical ESR
Ceramic			
NPO/COG	0.5 pF–0.1 μF	Good temperature stability	0.12 Ω at 1 MHz for 0.1 μF surface mount capacitor
X7R/Y5R	1 pF–3.3 μF	Nonlinear variation with temperature	
Z5U/Y5U	0.001 pF–10 μF	Poor temperature and voltage stability	
Film			
Polypropylene	0.5 pF–0.1 μF	Good temperature stability	0.11 Ω at 1 MHz for 1 μF surface mount capacitor
Polystyrene	100 pF–0.1 μF	Best overall specifications	
Polycarbonate	0.001–10 μF	Average temperature stability	
Polyester	100 pF–10 μF	High-temperature coefficient, lowest cost	
Electrolytic			
Aluminum	0.1–2.0 μF	Good overall specifications, high ESR, and leakage current	0.6 Ω at 100 kHz for 100 μF capacitor
Tantalum	0.001–8000 μF	Best overall specifications, low ESR and leakage current, less temperature sensitivity than aluminum	0.12 Ω at 100 kHz for 100 μF capacitor

13.1.2 Inductor Characteristics

The inductor also has a self-resonant frequency. The inductor remains inductive below and starts to appear as a capacitor above this frequency. Here is a series equivalent circuit of the inductor (Fig. 13.4):

The formula for calculating the resonant frequency of an inductor is the same as for a capacitor.

Fig. 13.4 Inductor
equivalent circuit

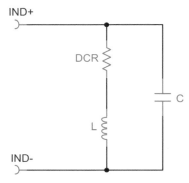

Fig. 13.5 Inductor
impedance response

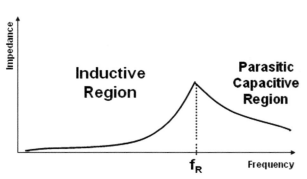

$$f_R = \frac{1}{2\pi\sqrt{LC}}.$$

Figure 13.5 shows the inductor frequency response. For a particular inductor, the impedance increases with frequency and reaches the highest impedance point at the resonant frequency, f_R. For frequencies above the resonant frequency, the impedance of the inductor is dominated by the parasitic capacitor, C, and this causes the impedance to decrease with frequency. It is recommended to operate in the inductive region of the curve as this region guarantees a close to ideal impedance response of the inductor. Like capacitors, there are different types of inductors and the two main ones are air core and magnetic core. Air core is the coil with air or insulating core and magnetic core is the coil wrapped around magnetic materials such as iron and ferrite. Inductors are commonly being used in RF and high-power circuits but are rarely being designed in high-speed DSP systems. Because it is better and lower cost to use ferrite beads to isolate and filter the noise in DSP systems.

Here are general rules for using inductors to filter noise in a DSP system:

- Inductors are expensive and are sensitive to noise. Depending on the switching speed of the signals propagating through it, an inductor can also generate and radiate noise.

- Inductors are commonly used to filter low-frequency noise in high current applications. In this case, designers need to add a high-frequency filter in series with the inductor to reject the high-frequency noise. Because inductors behave like a short circuit for noise with a frequency higher than the inductor resonant frequency.

13.1.3 Ferrite Bead Characteristics

Ferrite beads have electrical characteristics that are similar to ideal inductors. The key difference is that the ferrite bead has no or negligible parasitic capacitance until the frequency reaches GHz range as shown in Fig. 13.7. So, the ferrite bead behaves like an inductor over a wide frequency range. As shown in Fig. 13.6, ferrite bead always has a small DC resistance so review the specifications carefully and select a component that has the right AC impedance and low IR drop for the design. The ferrite bead generally performs well at frequencies higher than 30 MHz. It is commonly used to isolate power supplies and noise-sensitive circuits such clocks, video, and audio CODECs.

Some manufacturers provide free design tools to help engineers selecting and simulating the ferrite bead circuits. Figure 13.7 shows an impedance response provided by one of the ferrite bead design tools [3].

Two important parameters to select a ferrite bead are DC resistance and AC impedance at a given frequency. In general, assuming no issues with PCB space, it is best to select a device with the lowest DC resistance and highest impedance at the operating frequency. This yields the lowest IR drop across the ferrite bead while providing the highest noise rejection.

Fig. 13.6 Ferrite bead equivalent circuit

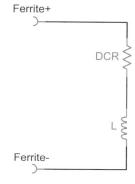

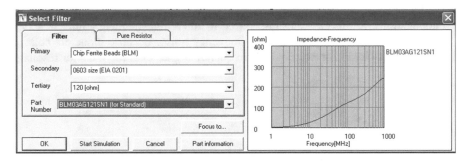

Fig. 13.7 Ferrite bead electrical characteristics

13.1.4   General Rule-of-Thumb Decoupling Method

The ideal way to decouple the supply noise is to have one capacitor between each of the power and ground pins of the DSP. Normally, this is physically not practical because the DSP package area is too small. So, designers must compromise by reducing the number of decoupling capacitors to fit in the general area underneath or above the DSP. Refer to the device data manual for a recommended method. But in general, here are the important considerations for decoupling:

- Add as many decoupling capacitors as space allows but do not put more capacitors than the DSP power pins.
- Add 8 bulk capacitors, 4 for Core and 4 for IO supplies. Place each bulk capacitor at each region of the DSP, with the region being defined as an edge or a corner of the DSP. Bulk capacitors act as a low-frequency noise filter and a charge storage device for the smaller decoupling capacitors. The use of four bulk capacitors is preferable to one large discrete component because this guarantees a shorter recharge path and a lower parasitic inductance path between the bulk and the decoupling capacitors.
- Keep in mind that all capacitors have equivalent series inductance (ESL) and equivalent series resistance (ESR). ESL and ESR reduce filtering effectiveness. So, select the smallest surface mount capacitors that can be used.

Figure 13.8 demonstrates a good scheme for decoupling a particular DSP. Refer to the device data manual to find more details. As shown in this figure, 0.01 μF ceramic capacitors are used for the decoupling capacitors and 10 μF tantalum capacitors are used as low-frequency filtering components. Typically, designers must go back and change the values to optimize them for their applications. A good approach is changing the capacitor values to achieve less than 50 mV power supply ripple for the IO rail and less than 20 mV for the Core rail. Another good rule is to use ceramic capacitors for high-frequency decoupling and tantalum capacitors for low-frequency filtering. This is because tantalum capacitors come in higher values than ceramic capacitors as shown in Table 13.1. These two types of capacitors

Fig. 13.8 General rules for decoupling

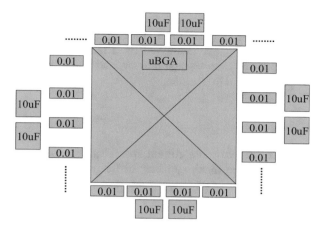

provide the low ESR and ESL which are needed for low noise and low EMI designs operating over a wide range of voltages, temperatures, and frequencies.

These general rules are only applied to the digital Core and IO power pins of the DSP. PLL and other analog power pins need to include better filtering schemes (Pi filters and or linear voltage regulators) to prevent the low-and high-frequency noise from affecting the performance of these circuits. Refer to the PLL (Chap. 6) and data converters (Chap. 7) for more details.

13.1.5   Analytical Method of Decoupling

Another method of decoupling power supply noise from a DSP system is calculating the total capacitance required to keep the power supply ripple under a certain limit. Similar to the general rules of decoupling, this method provides a starting value that must typically be optimized. The large ball grid array (BGA) package typically used for DSPs behaves like a PCB itself with long traces routing from the die out to the balls. These traces can generate interference and are susceptible to crosstalk, power supply droop, and other electrical noise. The asymmetry analytical decoupling technique begins by dividing the DSP into 4 regions and then decoupling each region separately. Providing fewer decoupling capacitors in the low-speed section leads to a uniform reduction in noise and electromagnetic radiation. The rules for this decoupling technique are:

13.1.5.1   Core Voltage Decoupling Steps

- Divide the DSP package into 4 regions by drawing 2 diagonal lines across the 4 corners of the DSP as shown in Fig. 13.10. Be sure to keep a group of signals together, for example, keeping all the DDR signals in one region. The boundaries do not have to be diagonally divided as shown in Fig. 13.10.

- Conservatively estimate the current consumption of the Core voltage in the region, $I_{CRegion}$, as shown in the equation below by taking the maximum device current, $I_{CoreMax}$, multiplied by 2 (adding 100% margin) divided by the total number of Core voltage pins, N, and multiplied by the number of Core voltage pins, M, within a region.

$$I_{CRegion} = \frac{2 \times I_{CoreMax}}{N} \times M. \tag{13.4}$$

If the maximum current specification is not available in the datasheet, then estimate the maximum current by multiplying the typical current by 2 as in Eq. (13.5).

$$I_{CRegion} = \frac{4 \times I_{CoreTyp}}{N} \times M. \tag{13.5}$$

- Calculate the total decoupling capacitance for the region by applying Eq. (13.7)

$$I_{CRegion} = C_{Core} \frac{dV_{Core}}{dt}, \tag{13.6}$$

$$C_{Core} = I_{CRegion} \frac{dt}{dV_{Core}}, \tag{13.7}$$

where dt is the fastest rise time in the region and dV is the maximum ripple allowed for the Core voltage, assuming 10 mV ripple.
- Now, calculate the total bulk capacitance for the region by multiplying the total decoupling capacitance by 40. The rule recommended for bulk capacitance is at least 10 times the total decoupling capacitance [4]. Use one bulk capacitor per region to minimize the parasitic inductance between the bulk and the decoupling capacitors.
- To figure out the number of decoupling capacitors, review the PCB area to see how many capacitors can be placed within 0.5 in. or 1.25 cm of the power supply pins. It is preferable to use smaller size capacitors in order to have more capacitors in a region. If the DSP package is a full grid array as shown in Fig. 13.9, escape all the signals uniformly out in 4 different directions (Northwest, Northeast, Southwest, and Southeast) to create two lanes across the package. Now, use these two lanes to place the capacitors near the DSP Core and IO power pins. To find the decoupling capacitor value, divide the total capacitance by the number of capacitors allowed for the region. It is good to select a capacitor with a self-resonant frequency equal to the maximum frequency of the particular region. For example, if the SDRAM port runs at 100 MHz, then add at least one capacitor with the resonant frequency of 100 MHz in this region. The other capacitors within each region should have the highest possible resonant frequency. This helps mitigate EMI over a wide frequency range.

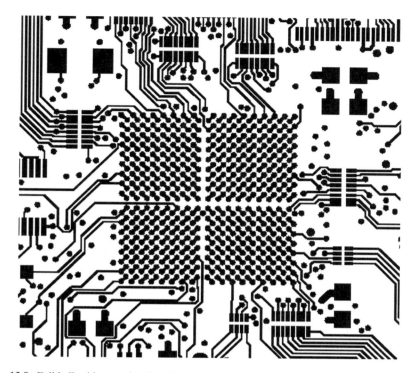

Fig. 13.9 Full ball grid array signal routings

13.1.5.2   IO Voltage Decoupling Steps

- Divide the DSP package into 4 regions by drawing 2 diagonal lines across the 4 corners of the DSP as shown in Fig. 13.10. Be sure to keep the signal groups together as indicated in Core voltage decoupling section.
- Count the number of IO voltage, inputs and outputs of each region.
- Conservatively estimate the IO current consumption of the DSP itself in the region, $I_{IORegion}$, as shown in Eq. (13.8) by taking the maximum device current specification, I_{IO}, divided by the total number of IO voltage pins, K, and multiplied by the number of IO voltage pins, J, within a region. Do not need to add margin to the IO current consumption here as the margin will be added in the next step.

$$I_{IORegion} = \frac{I_{IO}}{K} \times J. \tag{13.8}$$

- The total IO current is not equal to the IO current sourcing and sinking defined in the DSP datasheet. Most of the total IO current depends on the external loads, for example, resistive, capacitive, or transmission line. In this design, let us add a lot of margins by assuming a worst case scenario where all the IOs are outputs and

Fig. 13.10 Analytical decoupling technique

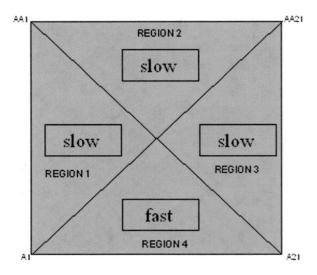

are loaded with transmission lines. In this case, each output current, $I_{IOTrans}$, is equal to the output voltage divided by the characteristic impedance of the transmission line, Z_o, as shown in Eq. (13.9).

$$I_{IOTrans} = \frac{V_{IO}}{Z_o}. \tag{13.9}$$

- In Eq. (13.10), the total IO current for the region is equal to the IO current of the DSP itself plus the IO current driving the transmission lines.

$$I_{IOTotal} = I_{IORegion} + J \times I_{IOTrans} \tag{13.10}$$

Substitute Eq. (13.9) into Eq. (13.10),

$$I_{IOTotal} = I_{IORegion} + J \times \frac{V_{IO}}{Z_o}. \tag{13.11}$$

- Calculate the total decoupling capacitance for the region by applying Eq. (13.13).

$$I_{IOTotal} = C_{IO} \frac{dV_{IO}}{dt}, \tag{13.12}$$

$$C_{IO} = I_{IOTotal} \frac{dt}{dV_{IO}}, \tag{13.13}$$

where dt is the fastest rise time in the region and dV is the maximum ripple allowed for the IO voltage, assuming 50 mV ripple.

- Now, calculate the total bulk capacitance for the region by multiplying the total decoupling capacitance by 40. The rule recommended for bulk capacitance is at least 10 times the total decoupling capacitance [4]. Use one bulk capacitor per region to minimize the parasitic inductance between the bulk and the decoupling capacitors.

- To figure out the number of decoupling capacitors, review the PC board area to see how many capacitors can be placed within 0.5 in. of the pins. If the DSP package being used is a full ball grid array, then apply the same technique outlined in the Core decoupling section to create two lanes for placing the decoupling capacitors nearby the power pins. To find the decoupling capacitor value, take the total capacitance value just calculated and divide it by the number of capacitors allowed for the region. It is good to select a capacitor with a self-resonant frequency equal to the maximum frequency of the particular region. For example, if the video port IO runs at 100 MHz, then add at least one capacitor with the resonant frequency of 100 MHz at this region. For the rest of the capacitors within that region, select the highest possible resonant frequency value.

This analytical decoupling method provides designers with a good starting point. As mentioned earlier, designers need to optimize the decoupling capacitors to ensure low noise and EMI during the board characterization process. The following example shows how this process can be applied to a typical design.

Example 13.1 Let us use a 289-pin BGA (Ball Grid Array) DSP [5]. Now, divide the 289-pin package into 4 regions by drawing two symmetry lines across the part as shown in Fig. 13.11. Then count the number of Core voltage pins, I/O voltage pins and signals, not including the ground pins, in each region. Also, pay special attention to the critical sections, such as external memory interface fast (EMIFF), phase-locked loop (PLL), and other high-speed serial/parallel ports. Assume all IOs outputs driving a 60-Ω transmission line and all the signal groups falling within a region. These are reasonable assumptions but there are cases where the boundaries of the regions had to be altered in order to keep the signal groups together. For the PLL and other analog power pins, the decoupling schemes are covered in Chap. 11.

The next step is to conservatively estimate the switching current requirements for each Region.

Table 13.2 shows the calculations of switching currents for all 4 Regions. The conservative assumptions used to calculate the capacitors in Table 13.2 are:

- Maximum Core current = Typical Current × 2 plus 100% margin = 170 mA × 2 × 2 = 680 mA.
- Device IO current = Typical IO current × 2 plus 100% margin = 45 mA × 2 × 2 = 180 mA.
- Total IO Current = Device IO Current plus IO Current Driving a Transmission Line.

Fig. 13.11 Bottom view of the DSP [5] package
REGION 1: 3 Core voltage pins, 8 I/O voltage pins, and 54 input/output pins.
REGION 2: 3 Core voltage pins, 4 I/O voltage pins, and 59 input/output pins.
REGION 3: 3 Core voltage pins, 3 I/O voltage pins, and 59 input/output pins check format of this section.
REGION 4: 4 Core voltage pins, 6 I/O voltage pins, and 55 input/output pins

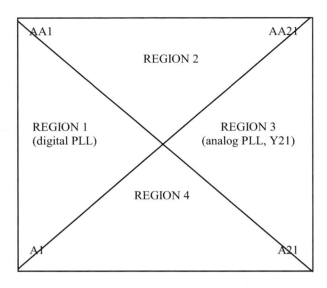

- Assuming that half of the inputs and outputs in the Region switching at the same time driving 60-Ω transmission lines, this is a conservative assumption since many of the signals in the 4 regions are too slow to be considered as transmission lines.

Since the Core and I/O voltage operate at different frequencies, they require separate decoupling calculations. The following shows the steps needed to calculate and select the decoupling capacitors for both Core and I/O supplies.

To find the decoupling capacitance, plug the peak current, the rise time, and the maximum ripple voltage parameters into Eq. (13.7) and solve for C. It is acceptable to assume that the maximum ripple voltage is 10 mV for Core and 50 mV for IO and the typical rise time is 2 nS.

$$C_{Core} = I_{CRegion} \frac{dt}{dV_{Core}}$$

Use the capacitor Eq. (13.13) to calculate the total capacitance for the IO voltage decoupling:

$$C_{IO} = I_{IOTotal} \frac{dt}{dV_{IO}}$$

Now let us calculate the total capacitance required for each region.

Table 13.2 Switching current estimation

Region	Total peak core current, $I_{IOTotal}$	Device IO current, $I_{IORegion}$	IO current for driving transmission lines, $I_{IOTrans}$	Total IO current, $I_{IORegion}$ plus $I_{IOTrans}$
Region 1	$\frac{680\ mA}{13} \times 3 = 157$ mA	$\frac{180\ mA}{21} \times 8 = 69$ mA	$\frac{3.3}{60} \times 54 = 2.97$ A	69 mA + 2.97 A = 3 A
Region 2	Same as 1, 157 mA	$\frac{180\ mA}{21} \times 4 = 34$ mA	$\frac{3.3}{60} \times 60 = 3.3$ A	3.3 A + 34 mA = 3.3 A
Region 3	Same as 1, 157 mA	$\frac{180\ mA}{21} \times 3 = 26$ mA	$\frac{3.3}{60} \times 59 = 3.3$ A	3.3 A + 26 mA = 3.3 A
Region 4	$\frac{680\ mA}{13} \times 4 = 209$ mA	$\frac{180\ mA}{21} \times 6 = 51$ mA	$\frac{3.3}{60} \times 55 = 3$ A	3 A + 51 mA = 3.1 A

Region 1

$$\text{Total Core capacitance,} \quad C_{\text{Core}} = 157 \ \text{mA} \frac{(2 \ \text{nS})}{(10 \ \text{mV})} = 0.03 \ \mu\text{F},$$

$$\text{Total I/O capacitance,} \quad C_{\text{IO}} = 3 \ \text{A} \frac{(2 \ \text{nS})}{(50 \ \text{mV})} = 0.08 \ \mu\text{F}.$$

There are 3 Core voltage pins operating at 150 MHz (CPU frequency) and 8 I/O voltage pins operating at 40 MHz (EMIFS frequency). It would be desirable to use multiple capacitors for the multiple supply pins, but there is a physical limitation due to the limited space available around the device. For the DSP [5] package, there is enough board space to place about 4 or 5 capacitors per region. In this case, select two capacitors with a total capacitance of around 0.03 μF. At least one of the capacitors should have a self-resonant frequency of around 150 MHz to decouple the Core voltage pins in region 1. Then, select three capacitors with a total capacitance of around 0.08 μF with at least one of the capacitors having the self-resonant frequency of around 75 MHz to decouple the I/O voltage pins in region 1.

In summary, for Core voltage in region 1, use two 0.022 μF (0.044 μF total) ceramic capacitors and, for the I/O voltage, use three 0.033 μF (0.099 μF total) ceramic capacitors.

The next step is calculating the bulk capacitors for both Core and IO. Bulk capacitor placement is not as critical as decoupling capacitor placement. But bulk capacitors are needed to filter the low-frequency ripple typically generated by switching power supply and to recharge the decoupling capacitors.

A rule-of-thumb is to select bulk capacitors with at least ten times the total decoupling capacitance. Let us use 40 times to be conservative. For the Core voltage,

$$40 \times \text{total Core capacitance} = 40 \times (0.03 \ \mu\text{F})$$
$$= 1.2 \ \mu\text{F for region 1 of the Core voltage,}$$

$$40 \times \text{total IO capacitance} = 40 \times 0.08 \ \mu\text{F} = 3.2 \ \mu\text{F for region 1 of the IO voltage.}$$

As mentioned earlier in this chapter, the best technique is adding 4 bulk capacitors to 4 regions of the DSP and the smallest tantalum capacitor available is 4.7 μF. In this case, select 4.7 μF tantalum bulk capacitors for both IO and Core voltages in region 1.

In summary, Fig. 13.12 shows the complete schematic diagram for decoupling Region 1 of the DSP. Next is to repeat the same steps for regions 2, 3, and 4.

Region 2

$$\text{Total Core capacitance,} \quad C_{\text{Core}} = 157 \ \text{mA} \frac{(2 \ \text{nS})}{(10 \ \text{mV})} = 0.03 \ \mu\text{F}$$

$$\text{Total I/O capacitance,} \quad C_{\text{IO}} = 3.3 \ \text{A} \frac{(2 \ \text{nS})}{(50 \ \text{mV})} = 0.13 \ \mu\text{F}.$$

Fig. 13.12 Region
1 decoupling capacitors

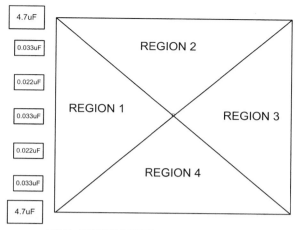

NOTE: SHADED COMPONENTS ARE FOR IO POWER AND
NON-SHADED COMPONENTS ARE FOR CORE POWER

   There are 3 Core voltage pins operating at 150 MHz (CPU frequency) and 4 I/O
voltage pins operating at 40 MHz (EMIFS frequency). For the DSP [5] package,
there is enough board space to place about 4 or 5 capacitors per region. In this case,
select two capacitors with a total capacitance of around 0.03 µF. At least one of the
capacitors should have a self-resonant frequency of around 150 MHz to decouple the
Core voltage pins in region 2. Then, select three capacitors with a total capacitance of
around 0.13 µF with at least one of the capacitors having the self-resonant frequency
of around 75 MHz to decouple the I/O voltage pins in region 2.
   In summary, for Core voltage in region 2, use two 0.022 µF (0.044 µF total)
ceramic capacitors and for the I/O voltage, use three 0.047 µF (0.14 µF total) ceramic
capacitors.
   The next step is calculating the bulk capacitors for both Core and IO. A rule-of-
thumb is to select bulk capacitors with at least ten times the total decoupling
capacitance. Let us use 40 times to be conservative.
   For the Core voltage,

$$40 \times \text{total Core capacitance} = 40 \times (0.03 \text{ µF}) = 1.2 \text{ µF for region 2.}$$

   For the IO voltage,

$$40 \times \text{total IO capacitance} = 40 \times 0.13 \text{ µF} = 5.2 \text{ µF for region 2 of the IO voltage.}$$

   In this case, select 4.7 µF tantalum capacitor for the Core voltage and 6.8 µF
tantalum capacitor for the IO voltage in region 2. Figure 13.13 shows the complete
decoupling schematic of region 2.

Fig. 13.13 Region
2 decoupling capacitors

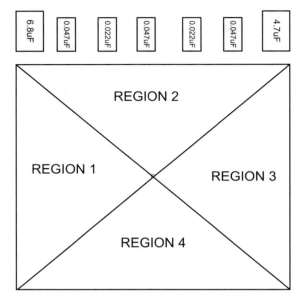

NOTE: SHADED COMPONENTS ARE FOR IO POWER AND
NON-SHADED COMPONENTS ARE FOR CORE POWER

Region 3

Region 3 has the same Core and IO currents as region 2. Therefore, the Core and IO
capacitors have the same values as the capacitors in region 2; there are two 0.022 µF
capacitors and three 0.047 µF capacitors. And for the bulk capacitors, one 4.7 µF
tantalum capacitor is for the Core voltage and one 6.8 µF tantalum capacitor is for the
IO voltage as shown in Fig. 13.14.

Region 4

$$\text{Total Core capacitance,} \quad C_{\text{Core}} = 209 \text{ mA} \frac{(2 \text{ nS})}{(10 \text{ mV})} = 0.042 \text{ µF,}$$

$$\text{Total I/O capacitance,} \quad C_{\text{IO}} = 3.1 \text{ A} \frac{(2 \text{ nS})}{(50 \text{ mV})} = 0.124 \text{ µF.}$$

There are 4 Core voltage pins operating at 150 MHz (CPU frequency) and 6 I/O
voltage pins operating at 40 MHz (EMIFS frequency). For the DSP [5] package,
there is enough board space to place about 4 or 5 capacitors per region. In this case,
select two capacitors with a total capacitance of around 0.042 µF. At least one of the

Fig. 13.14 Region 3 decoupling capacitors

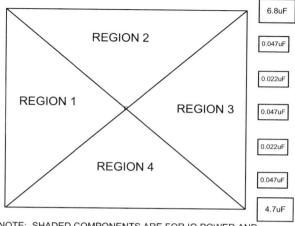

NOTE: SHADED COMPONENTS ARE FOR IO POWER AND
NON-SHADED COMPONENTS ARE FOR CORE POWER

capacitors should have a self-resonant frequency of around 150 MHz to decouple the Core voltage pins in region 4. Then, select three capacitors with a total capacitance of around 0.124 µF with at least one of the capacitors having the self-resonant frequency of around 75 MHz to decouple the I/O voltage pins in region 4.

In summary, for Core voltage in region 4, use two 0.027 µF (0.054 µF total) ceramic capacitors and for the I/O voltage, use three 0.047 µF (0.14 µF total) ceramic capacitors.

The next step is calculating the bulk capacitors for both Core and IO. A rule-of-thumb is to select bulk capacitors with at least ten times the total decoupling capacitance. Let us use 40 times to be conservative.

For the Core voltage,

$$40 \times \text{total Core capacitance} = 40 \times (0.054 \, \mu\text{F}) = 2.16 \, \mu\text{F for region 4.}$$

For the IO voltage,

$$40 \times \text{total IO capacitance} = 40 \times 0.14 \, \mu\text{F}$$
$$= 5.64 \, \mu\text{F for region 4 of the IO voltage.}$$

In this case, select 4.7 µF tantalum capacitor for the Core voltage and 6.8 µF tantalum capacitor for the IO voltage in region 4 as shown in Fig. 13.15.

Table 13.3 shows a summary of all the capacitors calculated for the 4 regions of the DSP and Fig. 13.16 shows the complete schematic.

Fig. 13.15 Region
4 decoupling capacitors

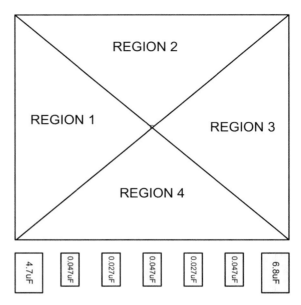

NOTE: SHADED COMPONENTS ARE FOR IO POWER AND
NON-SHADED COMPONENTS ARE FOR CORE POWER

Table 13.3 Summary of decoupling capacitors

Region	Ceramic caps for core (μF)	Bulk caps for core (μF)	Ceramic caps for IO (μF)	Bulk caps for IO (μF)
Region 1	2 × 0.022	1 × 4.7	3 × 0.033	1 × 4.7
Region 2	2 × 0.022	1 × 4.7	3 × 0.047	1 × 6.8
Region 3	2 × 0.022	1 × 4.7	3 × 0.047	1 × 6.8
Region 4	2 × 0.027	1 × 4.7	3 × 0.047	1 × 6.8

13.1.6   Target Impedance Method of Decoupling

Another method widely used in the industry is target impedance method which is the
maximum impedance allowed while maintaining the ripple voltage on a power
supply rail below the ripple specifications. The equation for target impedance, Z_{tar}, is

$$Z_{tar} = \frac{V_{dd} \times \text{Ripple}}{I_{transient}}, \tag{13.14}$$

where V_{dd} is the supply voltage rail, Ripple is the maximum percentage of the ripple
voltage, and $I_{transient}$ is the worst case transient current.

As discussed, the function of all the power supply decoupling capacitors is to
create a low AC current path to ground to prevent noise from affecting the system

Fig. 13.16 DSP decoupling schematic

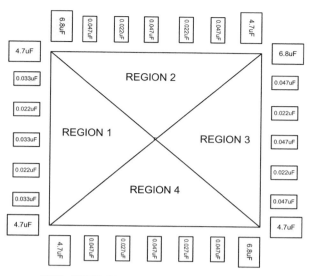

NOTE: SHADED COMPONENTS ARE FOR IO POWER AND
NON-SHADED COMPONENTS ARE FOR CORE POWER

performance. The impedance of the capacitors varies with frequencies, so it is critical to make sure the decoupling creates impedances that are lower than the target impedance of a particular power supply rail across the operating frequency. Use a power integrity tool like HyperLynx Decoupling Wizards [2] to simulate the target impedance as shown in Fig. 13.17. This example shows the target impedance is 40 mΩ from 100 Hz to 150 MHz and all the impedance curves (green, purple, and pink) are lower than the target impedance as designed.

13.1.7   Placing Decoupling Capacitors

It is important to place all the decoupling capacitors as close as possible to the pins, no more than 0.25 in. in most cases. The bulk capacitors should be placed as close as possible to the decoupling capacitors. This reduces the trace lengths, reducing the current loops and in turn lowering radiation while minimizing parasitic inductance. The best strategy is placing the decoupling capacitors on the bottom of the PCB right underneath the device being decoupled, and the bulk capacitors on top or bottom of the PCB close to the decoupling capacitors. Also, using via-in-pads to connect the capacitors to ground and power planes is the best way to keep parasitic inductances as low as possible. Due to costs, via-in-pad method has rarely been implemented in commercial products.

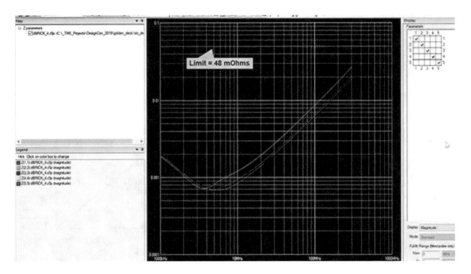

Fig. 13.17   Target impedance plot

In summary, there should be two bulk capacitors per region, one for Core and one for IO, and as many decoupling capacitors as space allows. Figure 13.18 shows a good example of the capacitor's placement on the bottom side of the PCB. The Core decoupling capacitors and four large bulk capacitors are placed on the interior of the BGA package in the open space right under the DSP. The IO decoupling and bulk capacitors are placed on the perimeter of the BGA package. This is possible because this particular BGA package is not a full BGA package where all the balls are fully populated on the bottom of the package.

If the package being used is a full BGA package, then it is necessary to route the signals from the DSP package out to external circuits as demonstrated in Fig. 13.19 that creates two lanes underneath the DSP for decoupling capacitors. Now, use these two lanes and populate as many decoupling capacitors as the lanes allow. In this case, the bulk and some of the IO capacitors can be placed on the perimeter of the DSP package. The recommended rules for creating the lanes are as follows:

- Power and ground pins need to be closest to the lanes. This allows the shortest connection paths to the capacitors.
- The lanes do not have to be symmetrical as shown in Fig. 13.19.
- Designers can replace the lanes with capacitor islands underneath the DSP if this allows placing the decoupling capacitors near the power pins. For example, instead of having two lanes, designers may want to create many islands and each island can hold one or more decoupling capacitors.

Core Caps

IO Caps

Fig. 13.18 Good decoupling capacitors placement

13.2 High-Frequency Noise Isolation

The decoupling methods described up to now filter noise locally at the DSP. There are cases where the whole power supply plane for some critical sections needs to be isolated. This may be required to prevent external noise from entering these sections or to prevent noisy circuits such as oscillators from coupling onto the power plane. The power supply plane is generally isolated using either Pi or T filters. A Pi filter is constructed with two capacitors and one ferrite bead while a T filter requires one capacitor and two ferrite beads. Each of these filters is commonly used in series with the signals exiting and entering the system or the power supply to reduce the radiated emissions. The pass-band of the filter must be calculated precisely to ensure that the bandwidth is wide enough to pass the desired signals without degrading signal quality, especially critical parameters such as rise and fall times and amplitude.

13.2.1 Pi Filter Design

The filter bandwidth is calculated as follows:

For Pi filter in Fig. 13.20, starting from the DSP output, the first parallel component is C_1, second series component is Z, and third parallel component is C_2. Therefore, the bandwidth of this filter is determined by the three poles formed by C_1, ferrite bead, and C_2 assuming that the output impedance of the DSP is matched with the load impedance and is equal to (Fig. 13.21)

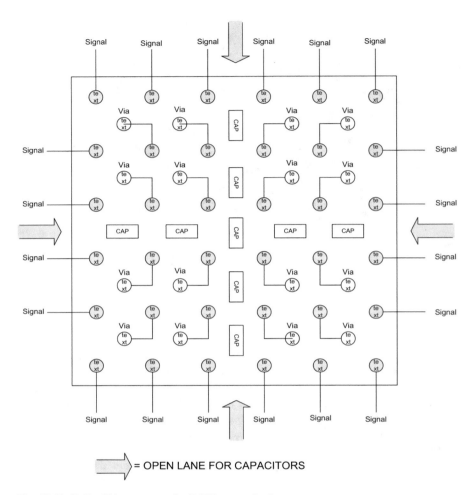

Fig. 13.19 Full grid layout example (DSP bottom view)

$$\sqrt{\frac{L_z}{2 \times C_1}}, \tag{13.15}$$

where L_z is the inductance of the ferrite bead Z.

For this special Pi filter assuming C_1 equal to C_2, the corner frequency of the 3-pole filter [6] is

$$f_C = \frac{1}{\pi \sqrt{L_z C_1}}. \tag{13.16}$$

Figure 13.22 shows the frequency response of this special Pi filter.

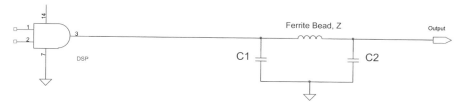

Fig. 13.20 Pi filter circuit for high-speed signals

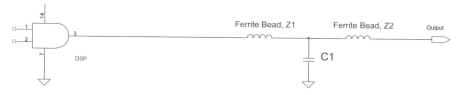

Fig. 13.21 Pi filter circuit for power supply isolation

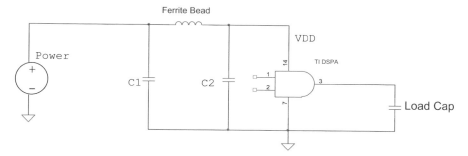

Fig. 13.22 Pi filter frequency response

13.2.1.1 Pi Filter Design Example

Let us design a Pi filter for a graphic controller's Red, Green, and Blue (RGB) analog signal outputs driving a computer monitor. Assuming that RGB signals have a 100 MHz analog bandwidth, calculate the filter components as follows.

Let $f_C = 200$ MHz. Setting the filter at 200 MHz provides a 100 MHz margin (200 MHz minus 100 MHz) to make sure that the filter is not affecting the video signal bandwidth.

The filter corner frequency is

$$f_C = \frac{1}{\pi\sqrt{L_z C_1}} = 200 \text{ MHz.}$$

Let us select a ferrite bead with 100 Ω impedance at 100 MHz and calculate L_z. The impedance, Z, of the ferrite bead is

$$Z = 2\pi fL = 100\ \Omega,$$

$$L_z = \frac{100}{2\pi f} = \frac{100}{2\pi(100 \times 10^6)} = 0.16\ \mu\text{H}.$$

Now, calculate C_1 by substituting L_z and f_C into the Eq. (13.16) and solve for C_1.

$$f_C = \frac{1}{\pi\sqrt{L_z C_1}} = 200\ \text{MHz},$$

$$200 \times 10^6 = \frac{1}{\pi\sqrt{(0.16 \times 10^{-6})C_1}},$$

$$C_1 = 15.8\ \text{pF}.$$

Therefore, the Pi filter has two 15.8 pF capacitors and one 0.16 μH inductor. Now, let us use an analog circuit simulator [7] to verify the design.

To match the source and load impedance, use Eq. (13.15) and calculate R_1 and R_2 values. In this case, $R_1 = R_2 = 71\ \Omega$ for $C_1 = 15.8$ pF and $L_z = 0.16$ μH.

Figure 13.23 shows the simulation results of the circuit model shown in Fig. 13.24. In the pass-band from DC to 100 MHz, the circuit shows a −6 dB attenuation. This is because the voltage divider is formed by the 71-Ω source resistor and 71-Ω load resistor. In this case, the attenuation is

$$= 20 \log_{10} \frac{V_{F2}}{V_{F1}}, \tag{13.17}$$

where

$$V_{F2} = \frac{71}{71 + 71} V_{F1}. \tag{13.18}$$

Now, substitute Eq. (13.18) into Eq. (13.17) and solve to calculate the attenuation.

$$\text{Attenuation} = 20 \log_{10} \frac{71}{71 + 71} = -6\ \text{dB}.$$

This correlates with the simulation results showing a −6 dB signal attenuation within the pass-band.

For the filter corner frequency, the simulation results show that the signal starts rolling off at 100 MHz with a slope of 60 dB/decade. This is correct as this is a 3-pole lowpass filter and each pole has a 20 dB/decade slope. Since the −3 dB corner

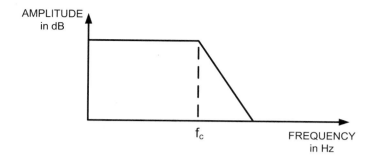

Fig. 13.23 Pi filter circuit simulation results

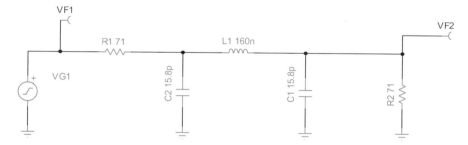

Fig. 13.24 Pi filter circuit model for simulation

frequency of each pole is at 200 MHz and the Pi filter has 3 poles (2 capacitors and 1 ferrite bead), the combined corner frequency at 200 MHz has a -9 dB attenuation as shown in the simulation.

13.2.2 T Filter Design

The filter bandwidth is calculated as follows:

For the T filter in Fig. 13.25, starting from the DSP output, the first series component is L_1, second parallel component is C_1, and third series component is L_2. Therefore, the bandwidth of this filter is determined by the three poles formed by the two equal inductance ferrite beads and C_1 assuming that the output impedance of the DSP is matched with the load impedance and is equal to

$$\sqrt{\frac{L_z}{2 \times C_1}},\tag{13.19}$$

where L_z is the inductance of the ferrite bead Z.

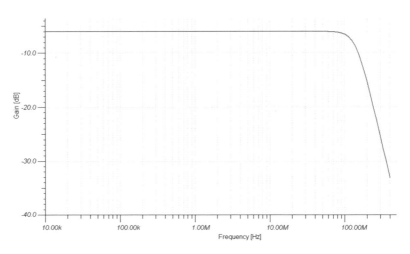

Fig. 13.25   T filter circuit for high-speed signals

For this special T filter assuming L_1 equal to L_2, the corner frequency of the 3-pole filter [6] is

$$f_C = \frac{1}{\pi\sqrt{L_z C_1}}. \qquad (13.20)$$

Figure 13.26 shows the frequency response of this special Pi filter.

13.2.2.1   T Filter Design Example

Let us design a T filter for a graphic controller's Red, Green, and Blue (RGB) analog signal outputs driving a computer monitor. Assuming that RGB signals have a 100 MHz analog bandwidth, calculate the filter components as follows:

Let $f_C = 200$ MHz. Setting the filter at 200 MHz provides a 100 MHz margin (200 MHz minus 100 MHz) to make sure that the filter is not affecting the video signal bandwidth.

The filter corner frequency is

$$f_C = \frac{1}{\pi\sqrt{L_z C_1}} = 200 \text{ MHz}.$$

Let us select a ferrite bead with 100 Ω impedance at 100 MHz and calculate L_z. The impedance, Z, of the ferrite bead is

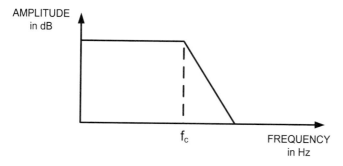

Fig. 13.26   T filter frequency response

$$Z = 2\pi f L = 100 \ \Omega,$$

$$L_z = \frac{100}{2\pi f} = \frac{100}{2\pi (100 \times 10^6)} = 0.16 \ \mu H.$$

Now, calculate C_1 by substituting L_z and f_C into the Eq. (13.20) and solve for C_1.

$$f_C = \frac{1}{\pi \sqrt{L_z C_1}} = 200 \ \text{MHz},$$

$$200 \times 10^6 = \frac{1}{\pi \sqrt{(0.16 \times 10^{-6}) C_1}},$$

$$C_1 = 15.8 \ \text{pF}.$$

Therefore, the T filter has two 0.16 μH ferrite beads and one 15.8 pF capacitor. Now, let us use an analog circuit simulator [7] to verify the design.

To match the source and load impedance, use Eq. (13.18) and calculate R_1 and R_2 values. In this case, $R_1 = R_2 = 71 \ \Omega$ for $C_1 = 15.8 \ \text{pF}$ and $L_1 = L_2 = 0.16 \ \mu H$.

Figure 13.27 shows the simulation results of the circuit model shown in Fig. 13.28. In the pass-band from DC to 100 MHz, the circuit shows a −6 dB attenuation. This is because the voltage divider is formed by the 71-Ω source resistor and 71 Ω load resistor. In this case, the attenuation is

$$= 20 \log_{10} \frac{V_{F2}}{V_{F1}}, \tag{13.21}$$

where

$$V_{F2} = \frac{71}{71 + 71} V_{F1}. \tag{13.22}$$

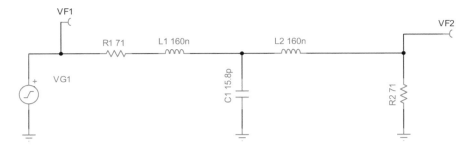

Fig. 13.27 Pi filter circuit simulation results

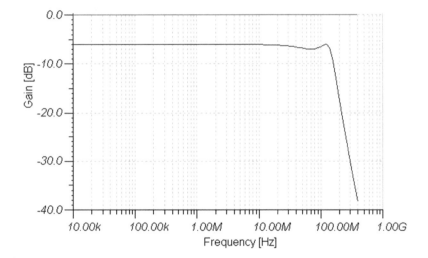

Fig. 13.28 T filter circuit model for simulation

Now, substitute Eq. (13.22) into Eq. (13.21) and solve to calculate the attenuation.

$$\text{Attenuation} = 20 \log_{10} \frac{71}{71 + 71} = -6 \text{ dB}.$$

This correlates with the simulation results showing a -6 dB signal attenuation within the pass-band.

For the filter corner frequency, the simulation results show that the signal starts rolling off at 100 MHz with a slope of 60 dB/decade. This is correct as this is a 3-pole lowpass filter and each pole has a 20 dB/decade slope. Since the -3 dB corner frequency of each pole is at 200 MHz and the T filter has 3 poles (2 ferrite beads and 1 capacitor), the combined corner frequency at 200 MHz has a -9 dB attenuation as shown in the simulation.

Table 13.4 shows a comparison between T and Pi filters.

Table 13.4 Pi and T filters comparison

	Pi filter	T filter
Components required	Two capacitors and one ferrite	Two ferrites and one capacitor
Cost	Slightly lower, capacitors are less expensive	Slightly higher
Effectiveness	Slightly better because having a capacitor to ground next to the connector reduces RF current loop areas	Slightly lower, larger RF current loop areas
Power supply isolation	Recommended for power supply filtering because the capacitor can be placed close to the power supply pin	Not as effective as the Pi filter, the capacitor cannot be placed close to the pin

13.3 Summary

Power supply decoupling and noise isolation techniques discussed in this chapter provide practical design considerations and theoretical approaches to design the optimum filters for noise isolation. Poor decoupling techniques are the number one root cause of random DSP system failures. So, to improve the probability of design success and to prevent random logic delete/remove line spacing failures caused by excessive system noise, designers need to do the following:

- Apply the General Rules for decoupling or the Analytic Decoupling methods described in this chapter.
- Use power integrity tool to simulate decoupling capacitors to make sure that the decoupling method provides good low impedance paths to ground, less than target impedance.
- Follow the guidelines shown in this chapter to place the decoupling capacitors properly. Select the right components (ferrite beads, inductors, capacitors, or resistors) for the design.
- Use Pi filters to isolate noise and place the capacitor as close to the connector as possible. This minimizes the RF current loops while providing some ESD protection.
- Use an analog simulator [7] and simulate the design to verify all the calculations before going into layout.

References

1. H. Johnson, M. Graham, *High-Speed Digital Design—A Handbook of Black Magic* (Prentice Hall PTR, Upper Saddle River, NJ, 1993)
2. Mentor Graphics, HyperLynx Signal Integrity Simulation Software (2004). http://www.mentor.com/products/pcb-system-design/circuit-simulation/hyperlynx-signal-integrity/
3. Murata Manufacturing Co., Murata EMI Filter Selection Simulator (2009)

4. O. Henry, *Electromagnetic Compatibility Engineering* (John Wiley and Sons, Hoboken, NJ, 2009)
5. Texas Instruments Inc., OMAP5910 Dual-Core Processor Data Manual (2002). http://focus.ti. com/lit/ds/symlink/sm320c6713b-ep.pdf
6. K. Kenneth, *Electromagnetic Compatibility Handbook* (CRC Press, Boca Raton, FL, 2005)
7. Texas Instruments Inc., Spice-Based Analog Simulation Program (2008). http://focus.ti.com/ docs/toolsw/folders/print/tina-ti.html

Chapter 14
Printed Circuit Board (PCB) Layout

Once all the circuits have been designed and simulated, the next step is board layout. This is one of the most critical steps in the development process because the effectiveness of the circuits designed depends on where the components are placed relative to the DSP/CPU pins and the traces that are being routed to those pins. Also, the board layout has a big effect on noise, crosstalk, and transmission line effects so optimizing the layout can minimize these effects. This chapter covers the printed circuit board stackup, routing guidelines, and layout techniques for low noise and EMI.

14.1 Printed Circuit Board (PCB) Stackup

First, designers must determine the minimum number of PCB layers and then configure the board stackup. Here are some general guidelines:

- Perform layout experiments and refer to the DSP/CPU reference design package to find the minimum number of layers required to escape the signals out from the device's package. Typically, DDR layout guidelines dictate the number of PCB layers to allow for escaping all the signals out from the DSP/CPU.
- Consider the need for high-speed signals to be shielded between the ground and power planes.
- Are there buses, such as USB, Ethernet, and RapidIO, that require a tight differential impedance specification? If so, designers need to follow the industry guidelines to control the differential impedance of these buses.
- Does the PCB manufacture require a certain trace width and spacing? This determines whether a trace can be routed between the balls of a small pitch BGA package. For good signal integrity with minimum skin effect losses, keep the trace width between 4 and 12 mils. A common choice is a 5-mil trace and 5-mil spacing.

© The Author(s), under exclusive license to Springer Nature Switzerland AG 2023
T. T. Tran, *High-Speed System and Analog Input/Output Design*,
https://doi.org/10.1007/978-3-031-04954-5_14

Fig. 14.1 Adjacent power and ground board stackup

- Is one power plane and one ground plane sufficient?
- Does the DSP system require a controlled impedance board? This is more expensive but allows the board to be optimized from a signal integrity standpoint.

Two PCB stackup topologies are commonly used, adjacent power and non-adjacent power and ground planes. Figure 14.1 [1] shows what can or cannot be done with each layer when the design is implemented on a 6-layer PCB for the adjacent power and ground topology.

The parallel capacitance, C_{pp}, between the power and ground planes is calculated as

$$C_{pp} = k \frac{\varepsilon_r A}{d} \text{pF}, \tag{14.1}$$

where k is 0.00885 m, ε_r, dielectric constant = 4.1–4.7 for FR4 type PCB, A is area of the power and ground planes in mm^2, and d is the distance between the power and ground planes in mm.

When using this topology, designers need to consider these points:

- As shown in the equation for C_{pp}, the distance, d, between the power and ground planes determine the board capacitance. Reducing the distance increases the capacitance and reduces high-frequency impedance. The limiting factor is how closely the layers can be packed together while still maintaining the quality and reliability of the design. Refer to the specifications from PCB manufacturers to understand the minimum requirements between the layers.
- Route the high-speed signals on the planes next to the power and ground planes. If possible, route all the high-speed signals next to the ground plane. If not, then it is also acceptable to route them next to the power plane.
- In Fig. 14.1, the best routing layer is Layer 2 because it is next to the ground plane. This provides optimal current return paths which help reduce radiation. Therefore, the adjacent power and ground topology is recommended for DSP systems operating at high frequency.

NON-ADJACENT P/G

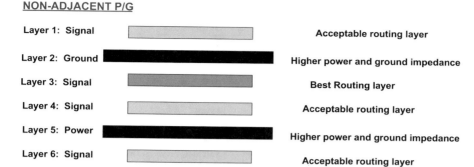

Layer 1: Signal		Acceptable routing layer
Layer 2: Ground		Higher power and ground impedance
Layer 3: Signal		Best Routing layer
Layer 4: Signal		Acceptable routing layer
Layer 5: Power		Higher power and ground impedance
Layer 6: Signal		Acceptable routing layer

Fig. 14.2 Non-adjacent power and ground board stackup

- The adjacent power and ground topology are not useable for DSP systems that require many layers to route the signals out from the DSP and interface with other circuits.

Figure 14.2 [1] shows a typical PCB stackup for the non-adjacent power and ground topology. The power and ground planes are placed in Layer 5 and Layer 2, respectively. Layer 3 is best for routing high-speed traces while Layer 1, Layer 4, and Layer 6 are acceptable. As shown in the figure, each of the routing layers is next to either a ground or power plane. Layer 3 is best because it is not only next to a ground plane but is also guarded by a power plane below it. This scheme is best for difficult-to-route DSP systems that do not operate at high frequency. One thing to keep in mind is that board capacitance becomes important for systems operating above 300 MHz [2].

Here are rules for doing non-adjacent board stackup design:

- For non-adjacent topology, the board capacitance, as shown in capacitance equation C_{pp}, is low and the board impedance is high between the power and ground planes. This is the opposite of what is needed for systems to have low noise and low EMI.
- This topology requires more high-frequency decoupling capacitors to compensate for the board characteristics.

14.2   Microstrip and Stripline

Table 14.1 shows the advantages and disadvantages of the two main signal routing technologies, Microstrip is shown in Fig. 14.3 where H is the height, W is the width of the signal trace, $W1$ is the width including the outside edge, T is the thickness of the trace, and ε_r is the dielectric constant. The Stripline is shown in Fig. 14.4.

Designers generally make compromises by using both topologies where some of the critical signals are routed between the ground and power planes.

Table 14.1 Microstrip and stripline comparison

	Microstrip topology	Stripline topology
1. Number of PCB layers	No special requirements	Requires signals to route between the ground planes so it is more expensive.
2. Routability	Easy and can route with minimum number of vias	Difficult to route with limited number of PCB layers. Also, vias are required which can cause signal quality degradation.
3. Signal quality	Acceptable	Good
4. EMI	Acceptable but an image plane is needed just below the routing layer.	Good because high-speed signals are shielded between the planes.

Fig. 14.3 Microstrip topology

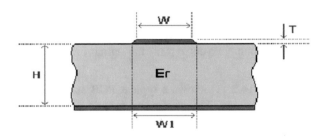

Fig. 14.4 Stripline topology

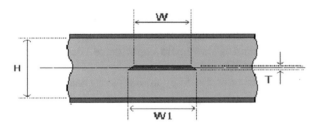

14.3   PCB Traces and Vias

14.3.1   PCB Trace Inductance

Trace inductance must be considered in high-speed design in order to optimize signal and power integrity. High inductance trace can cause excessive power supply droop as described in Chap. 13. Trace inductance can be calculated using Eq. (14.2) [3].

In Fig. 14.5, the trace inductance equation is

$$\text{Inductance} = 0.0002L\left(\ln \frac{2L}{(W+T)} + 0.2235 \frac{(W+T)}{L} + 0.5 \right) \mu\text{H}, \qquad (14.2)$$

where L is the trace length, W is the width, and T is the thickness. All dimensions are in mm.

Fig. 14.5 Trace structure

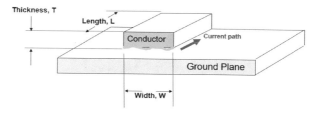

14.3.2   PCB Trace Impedance

For DC, the current flows through the entire area of the trace and the DC resistance is

$$Rdc = L/(\sigma WT), \tag{14.3}$$

where L is the length of the trace, σ is the conductivity of the metal, W is the width of the trace, and T is the thickness of the trace.

For AC, the impedance is

$$Rac = L/(\sigma w \delta), \tag{14.4}$$

where L is the length, σ (sigma) is the conductivity, W is the width, and δ is the skin depth. Skin depth is small compared to the thickness of the trace and is shown in Eq. (14.5).

$$\delta(delta) = \sqrt{1/(\pi f \mu \sigma)}, \tag{14.5}$$

where f is the frequency, μ (Mu) is the permeability of free space ($4\pi \times 10 - 7$ H/m), and σ is the conductivity (copper $= 5.6 \times 10^7$). Or

$$\delta = 66 \times 10^{-6} \ \mathbf{m}\sqrt{1/f}, \tag{14.6}$$

where **m** is the unit in meters.

For low frequency, the current distributes evenly on the conductor while for frequency higher than 10 MHz, the current concentrates on the surface of the conductor, and the depth of the current concentration is the skin depth as shown in Fig. 14.6. In general, DC resistance and AC impedance must be sized appropriately to guarantee good reliability and signal integrity. DC resistance is guarded by IPC-2221 industry standard, and there are many free trace calculators on the web, for example [4].

Design Example 14.1: Trace Width Calculations
Given:

Trace thickness = 1.4 mils (1.4/1000 in.),
Trace length = 5.8 in.,
Temperature = 55 °C, and
Current = 2 A

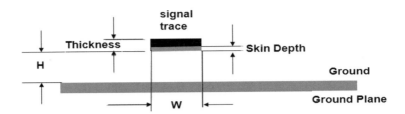

Fig. 14.6 Trace impedance

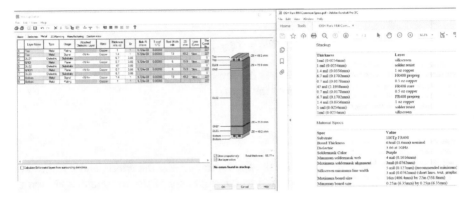

Fig. 14.7 PCB stackup

What is the minimum external Trace Width to guarantee that the design is IPC-2221 compliant? The trace can be a microstrip or a stripline, depending on a particular design.

Answer: Using PCB Trace Width Calculator [4], the answer is 15.5 mils minimum width. In this case, the voltage drop is 0.398 V which may be out of the range required by the receiver. Keep in mind that the IPC 2221 only gives a reliability limit, it is up to the engineers to determine whether the voltage drop is acceptable.

Design Example 14.2: Characteristic Impedance Calculations
Given:

A 4-layer PCB layout in Fig. 14.7 with two middle ground layers,
External trace thickness = 1.4 mils,
Height (from trace to the adjacent ground) = 6.7 mils,
Trace width = 13 mils, and
Dielectric constant (prepreg material) = 3.66

Calculate the characteristic impedance of the trace.

Answer: Using Microstrip Impedance Calculator [5] below, the answer is 51 Ω.

Inputs

Trace Thickness	1.4	mil	∨
Substrate Height	6.7	mil	∨
Trace Width	13	mil	∨
Substrate Dielectric	3.66		

Calculate

Output

Impedance (Z):	51.0	Ohms

14.3.3   Vias

Vias, Fig. 14.8, have both inductance and capacitance components [3] that can cause signal and power integrity issues. Vias need to be carefully designed to minimize parasitics and verified using 3D EM Advanced Solver [6].

Fig. 14.8 Via structure

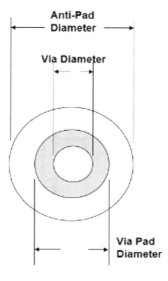

$$\text{Via Inductance} = 2T\left(\ln\frac{4T}{D0} + 1\right)\text{nH}, \tag{14.7}$$

where T is the thickness of the board and $D0$ is the via diameter, all in centimeters.

$$\text{Via Capacitance} = \left(\frac{0.55 \times \varepsilon_r \times T \times D1}{D2 - D1}\right)\text{pF}, \tag{14.8}$$

where ε_r is the dielectric constant of the board, T is the thickness of the board (via depth), $D1$ is the via pad diameter, and $D2$ is the anti-pad diameter (void area). All dimensions are in centimeters.

The parasitics of vias and traces must be included in the system simulations to get accurate results, and the method is to use an Advanced 3D Solver [6] tool to create s-parameter models for the simulations.

14.4   Image Plane

Image plane concept demonstrated in [7] is defined as having a ground or power plane right underneath the signal routing layer. This provides a shielding layer which greatly reduces the radiated emissions as shown in Fig. 14.9. The system failed EMI initially when there was no image plane inserted in the PCB stackup. The system passed EMI limits when the image plane was included in the PCB stackup as shown in Fig. 14.10.

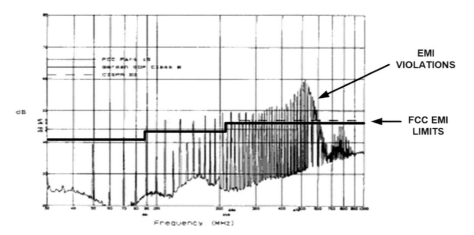

Radiated Emissions from PCB

Fig. 14.9 PCB radiated emissions without image plane

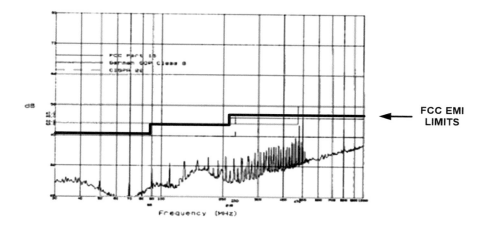

Radiated Emissions from PCB with Image Plane

Fig. 14.10 PCB radiated emissions with image plane

14.5 PCB Routing Guidelines

In high-speed system design, PCB routings and component placements must be carefully done and simulated, using signal and power integrity tools. Here is a list of the most important guidelines for routing.

- High-speed traces must have the right target characteristic impedance.
- Single ended traces need to be separated by at least three times the width. This spacing depends on the speed and the overlapping distance. Must do crosstalk simulations to confirm the spacings.
- Differential pair must be routed together, and the spacing between the positive and negative signals needs to be at least 1.5 times the width of the trace.
- The spacing between two differential pairs needs to be at least five times the width of the trace.
- If the differential pair needs to be transitioned from one layer to another layer, two ground vias must be added next to the two signals within the pair. See Chap. 10 for a detailed design example.
- If delay needs to be added to one of the signals in one differential pair, only insert serpentines to add delay near the vias that have the mismatch as shown in Fig. 14.11. The recommendations [8] are: (1) $A = B = C = D$; (2) $E = F = G = 3 \times$ Width; (3) $S2 < 2S1$. And the length from A to D needs to be longer than 100 mils.

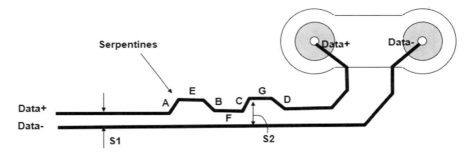

Fig. 14.11 Delay matching within one differential pair

14.6 Summary

PCB routing and board stackup are major contributors to design issues related to EMI, noise, signal integrity, and power integrity, so designers must apply best practices, including extensive simulations before releasing the design to fabrication, to improve the probability of having a first-pass success. The guidelines listed in Sect. 14.5 must be followed to ensure proper routings of single-end or differential signals.

Also, high-speed design requires having image planes in the PCB stackup to control the trace characteristic impedance as shown in Design Example 14.2. The use of an image plane, a ground or power plane located next to the routing layer, provides low inductance current return paths for high-speed signals. The image plane helps reduce current loop areas and minimizes the potential differences on the ground plane. Experiments conducted in [7] compare the EMI for PCBs with and without an image plane. They demonstrated that a PCB with an image plane shows around 15 dB reduction in EMI across the frequency spectrum.

References

1. M. Montrose, *Printed Circuit Board Design Techniques for EMC Compliance* (The Institute of Electrical and Electronics Engineers, New York, 2000)
2. Texas Instruments Inc., Design Guidelines: Integrated Circuit Design for Reduced EMI. Application Note (2000)
3. J. Ardizzoni, A Practical Guide to High-Speed Printed Circuit Board Layout, Analog Dialogue 39-09, Sept 2005
4. PCB Trace Width Calculator, https://www.4pcb.com/trace-width-calculator.html
5. Trace Resistance Calculator, https://www.allaboutcircuits.com/tools/trace-resistance-calculator/
6. Mentor Graphics (VX.2.7) HyperLynx Signal Integrity Simulation Software, http://www.mentor.com/products/pcb-system-design/circuit-simulation/hyperlynx-signal-integrity/
7. M. Montrose, *Analysis on the Effectiveness of Image Planes Within a Printed Circuit Board* (The Institute of Electrical and Electronics Engineers, New York, 1996)
8. Signal Integrity (SI) in High-Speed PCB Designs, Intel Corporation, ID: 683864, 18 Mar 2021

Chapter 15
Electromagnetic Interference (EMI)

Radiated emissions in high-speed systems are caused by fast switching currents and voltages propagating through printed circuit board traces. As system speed increases, printed circuit board traces are becoming more effective antennas, and these antennas are radiating unwanted energies that interfere with other circuitry and with other systems located nearby. This section outlines different ways to design for low EMI and find the root cause of EMI problems when they occur. It only covers the electrical design aspects of EMI even though shielding, cabling, and other mechanical fixes can also be used to help reduce the emissions below the maximum allowable limits. In general, mechanical solutions are expensive for high-volume designs. Even worse, the mechanical solutions may have to change when the system speed increases.

15.1 FCC Part 15 B Overview

To prevent systems from interfering with each other, the FCC sets maximum limits known as FCC Part 15 A for commercial products and FCC Part 15 B [1] for consumer devices as shown in Fig. 15.1.

The following lists some of the most common sources of EMI in high-speed DSPs:

- Fast switching digital signals such as clocks, memory buses, PWM (switching power supplies)
- Large current return loops
- Not having an adequate power supply decoupling scheme around large DSPs
- Transmission lines
- Printed circuit board layout and stackup, lack of power, and ground planes
- Unintentional circuit oscillations

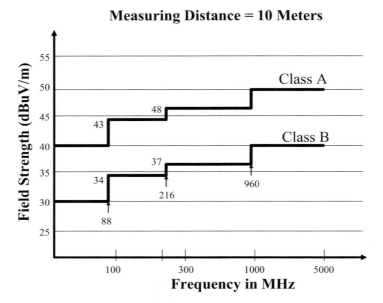

Fig. 15.1 FCC radiated emissions limits

15.2   EMI Fundamentals

The five main sources of radiation are digital signals propagating on traces, current return loop areas, inadequate power supply filtering or decoupling, transmission line effects, lack of power, and ground planes.

Radiation is classified into two modes, differential mode radiation and common mode radiation. It is important for engineers to understand the differences between the two modes in order to develop an effective scheme to mitigate the problem. In DSP systems, all electrical currents propagate from the source to the load and return to the original source. This mechanism generates a current loop which creates differential mode radiation as shown in Fig. 15.2 [2].

Differential mode radiation is directly related to the length of the signal trace, the driving current, and the operating frequency. The electric field caused by differential mode radiation is

$$E = 87.6 \times 10^{-16} \left[f^2 A I \right], \tag{15.1}$$

where f is the operating frequency; A is the current loop area created by the trace length and the board stackup; and I is the driving source current.

Common mode radiation is generated by a differential voltage between two points on a ground plane. It typically radiates from cables connected to the board or chassis. In theory, 100% of the source current returns to the source but a small portion of the current spreads over the entire plane before finding its way back to the source. This current creates an imbalance in the ground potential and causes common mode radiation as shown in Fig. 15.3 [2].

Fig. 15.2 Differential mode radiation

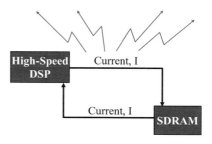

Fig. 15.3 Common mode radiation

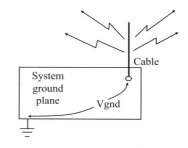

Fig. 15.4 Relationship between common and differential mode radiation

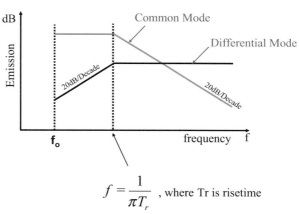

$$f = \frac{1}{\pi T_r} \quad , \text{ where Tr is risetime}$$

The electric field generated by common radiation is directly related to the frequency propagating on the cable, the length of the cable, and the current driving the cable. Here is an equation to calculate the common mode radiation in an open field.

$$E = 4.2 \times 10^{-7}[fLI], \qquad (15.2)$$

where f is the frequency, L is the length of the cable in meter, and I is the source current.

The relationship between the common mode and differential mode radiation for a given signal is shown in Fig. 15.4 [2]. In general, the differential mode dominates at a higher frequency spectrum while the common mode radiates more energy around the operating frequency.

15.3   Digital Signals

A digital or squarewave signal consists of a series of sine and cosine signals superimposed on one another. In the frequency domain, a squarewave consists of many higher frequency harmonics, and the harmonic radiated energy directly depends on the rise time and the pulse width of the signal as shown in Fig. 15.5.

In Fig. 15.5 assuming a 50% duty cycle signal where only odd harmonics are present, the amplitudes of the harmonics decay slowly as frequency increases. The first pole frequency is at

$$f_{-3\text{dB}} = \frac{1}{\pi P_{\text{W}}}, \tag{15.3}$$

and a second pole is at

$$f_{-3\text{dB}} = \frac{1}{\pi T_{\text{r}}}. \tag{15.4}$$

P_{W} and T_{r} are the width and the rise time of the signal, respectively. Therefore, increasing the rise time increases attenuation of the harmonics which leads to lower

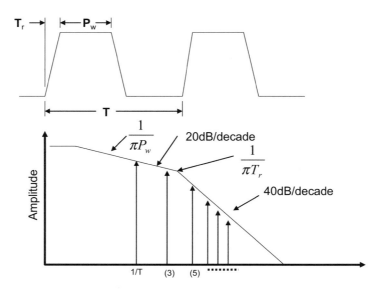

Fig. 15.5 Frequency spectrum of a squarewave

radiation. This method is not always practical because the slower rise time reduces the timing margin and may violate electrical requirements such as setup and hold times.

The best technique to minimize EMI generated by digital signals is keeping the high-speed signal traces as short as possible. It is a good practice for engineers to go through a design and analyze the traces to see if they are effective antennas or not. A good rule-of-thumb is keeping the length of the trace less than the wavelength (λ) divided by 20. Here is the equation

$$\text{max\_trace\_length} = \frac{\lambda}{20} = \frac{c}{20f}, \tag{15.5}$$

where C is the speed of light, 3×10^8 m/s, and f is the frequency.

For example, a 1.18-in. trace becomes an effective radiator when it is being driven by a 500 MHz signal. The 500 MHz signal is a fifth harmonic of a 100 MHz clock, common frequency in DSP systems today.

15.4 Current Loops

Current loops are the dominant sources of EMI, so it is important for designers to understand high-speed and low-speed current return paths and optimize the design to reduce the loop areas. Figure 15.6 shows two possible current return paths from points A and B; for high-speed current (>10 MHz), the return is right underneath the signal and, for low-speed current, the return is the shortest path back to the source.

Should high speed and low speed have hyphen in this figure and title?

As shown in Fig. 15.7 [3], current return creates a loop area that is directly related to the radiated electric field, so reducing the loop area lowers radiation. Skin effect modifies the current distribution within a conductor and increases resistance, so the high-speed current return is right underneath the signal. Skin effect is negligible at lower frequencies but increases as frequency rises. For a typical conductor used in

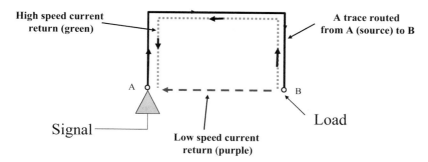

Fig. 15.6 High-speed and low-speed current loops

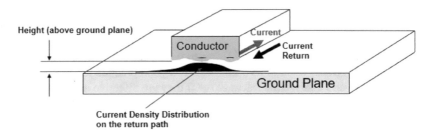

Fig. 15.7   High-speed current return on continuous ground plane

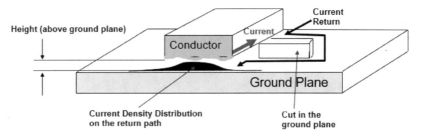

Fig. 15.8   High-speed current return on a discontinuous ground plane

DSP systems, a 10 MHz or higher trace is a high-speed signal. Providing a continuous ground plane right underneath a high-speed signal is the most effective way to achieve the lowest current loop area.

If the ground plane is not continuous underneath the high-speed signal, all crosstalk, reflections, and EMI will increase due to the impedance mismatch and larger current loop return area as shown in Fig. 15.8 [3].

15.5   Power Supply

The power supply is another major source of EMI because:

- The power supply is common to many high-speed sections in a design. RF signals may propagate from one section to another generating excessive EMI.
- A switching power supply generates fast current transients with a large amount of radiated energies. A 1 MHz switching power supply can radiate enough energy to fail EMI testing at the 100 MHz frequency range.
- Inadequate power supply decoupling may lead to excessive voltage transients on the power supply planes and traces.
- The power supply board layout can be a root cause of oscillations.

As shown in Fig. 15.9, decoupling the power supply reduces transients and provides a smaller current loop area. If the power supply trace in Fig. 15.9 is long

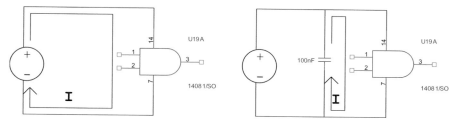

Fig. 15.9 Power supply decoupling reduces current loop area

and has no decoupling capacitor, the parasitic inductance is large and requires some time to charge up. This delay is the root cause of the power supply droop problem. Power supply droop occurs when the output buffer switches at a fast rate but is starved for the current needed to drive the load since the parasitic inductance between the power supply and the DSP becomes an open circuit.

Example 15.1

– A DSP BGA (ball grid array) package has a trace inductance of 1.44 nH.
– This output is driving a 3 in. trace with 1 nS rise time signal.
– This trace is being routed on a typical FR4 printed circuit board. Line characteristic impedance and IO voltage are 68 Ω and 3.3 V, respectively.

To estimate the power supply droop caused by the parasitic inductance, first let us estimate the peak current as follows. The dynamic IO current is the current transient for transmission line load, not steady-state resistive load.

$$I\,(\text{peak}) = \frac{\Delta V}{Zo} = \frac{3.3\ \text{V}}{68} = 48.5\ \text{mA}$$

Since the package inductance is 1.44 nH for 1 nS rise time signal, the internal voltage droop is

$$V\,(\text{droop}) = L\frac{dI}{dt} = (1.44\ \text{nH})\frac{48.5\ \text{mA}}{1\ \text{nS}} = 70\ \text{mV}$$

Typically, one DSP power supply pin is shared by many output buffers. This creates a larger droop and leads to higher radiation. This helps explain why good power supply decoupling is required for low EMI design.

15.6   Transmission Line

To combat TL effects, use simulation tools to fine-tune the series termination resistors to eliminate overshoots and undershoots caused by impedance mismatch explained in Chap. 6. Improving signal integrity design helps reduce EMI. But to minimize EMI, the series termination resistors should also be as large as possible without violating AC timing. A parallel termination resistor as shown in Fig. 15.10 is commonly used in RF and analog designs but is not practical for digital signals due to the amount of DC current drained by the 50-Ω resistor. If parallel termination is required, use a DC blocking capacitor in series with the resistor as demonstrated in Sect. 15.6.

Table 15.1 [4] shows the source current for different values of the series termination resistor. Changing the value from 10 to 39 Ω does not have much effect on the waveform [4], showing about 1 nS degradation, but dramatically reduces the source current which greatly lowers the radiated emissions. Figure 15.11 shows a DSP board with a 47-Ω series resistor added to the memory clock, reducing the radiated emissions 3 dB compared to the emissions of the signal without termination.

Overall, if slower rise time signals are acceptable and do not violate AC timing specifications, designers should use the largest resistor value to terminate high-speed signals to optimize the design from an EMI standpoint.

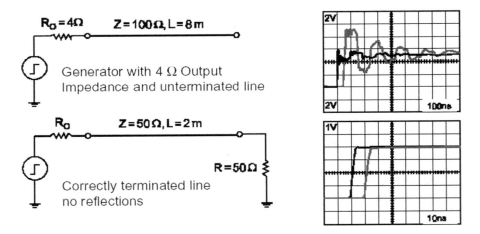

Fig. 15.10 Terminated and unterminated transmission lines

Table 15.1 Source current for different series termination

Series termination resistor value (Ω)	Peak source current (mA)
10	~40
22	~10
25	~5
30	~10
33	~9
39	~8

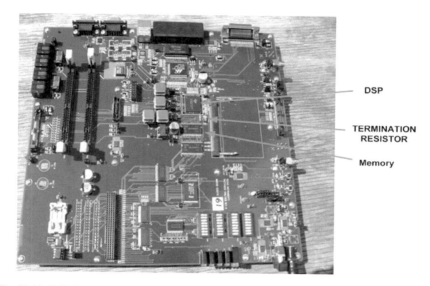

Fig. 15.11 DSP board with terminated clock

15.7 Power and Ground Planes

For high-speed DSP systems, it is getting more and more difficult to meet EMI regulations without using multiple layer PCBs and dedicating some of the layers as power and ground planes. Compared to a trace, a power or ground plane has a lower parasitic inductance and provides a shielding effect for high-speed signals. Power and ground planes also provide natural decoupling capacitance. As described in the PCB layout section of this document, natural decoupling capacitance occurs when power and ground planes are spaced closely, yielding higher capacitance. This effect becomes important at 300 MHz speed or higher. So, adding power and ground planes simplifies PCB routing and reduces the number of high-frequency decoupling capacitors required for the DSP.

Another important consideration for the PCB is layer assignment. Refer to the board layout chapter, Chap. 10, to determine the best board stackup for your application. Keep in mind that adding a ground plane directly underneath the high-speed signal plane creates an image plane that provides the shortest current return paths. Studies in [5, 6] show that image planes greatly reduce radiated emissions. The comparison between PCB with and without image plane is shown in Figs. 15.12 and 15.13.

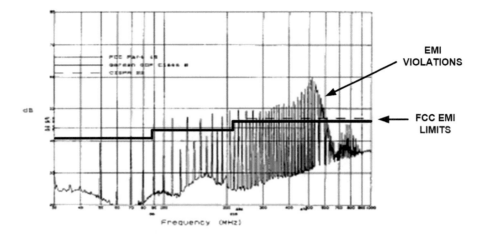

Radiated Emissions from PCB

Fig. 15.12   Radiated emissions without image plane

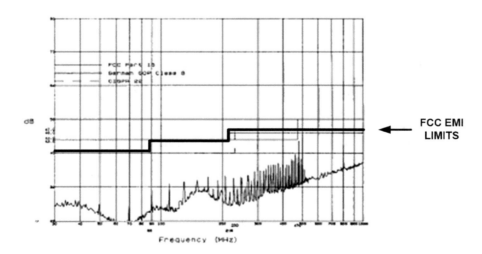

Radiated Emissions from PCB with Image Plane

Fig. 15.13   Radiated emissions with image plane

15.8   Summary: EMI Reduction Guidelines

In summary, here are the guidelines for low EMI system design:

- Add image planes wherever possible.
- Create ground planes if there are spaces available on the routing layers. Connect these ground areas to the ground plane with vias. Creating a quarter inch via grid is ideal.
- Add guard traces to high-speed signals if possible.
- Reduce the rise time of the signal if the timing is not critical. This can be accomplished by including series termination resistors on high-speed buses and fine-tuning the resistors for optimal signal integrity and EMI. Series termination resistors lower the source current, increase the signal rise time and reduce transmission effects. Substantial benefits can be achieved with this approach at a low cost.
- Keep the current loops as small as possible. Add as many decoupling capacitors as possible. Always apply current return rules to reduce loop areas.
- Keep high-speed signals away from other signals and especially away from input and output ports or connectors.
- Avoid isolating the ground plane. If this is required for performance reasons, such as with audio ADCs and DACs, apply current return rules to connect the grounds together.
- Avoid connecting the ground splits with a ferrite bead. At high frequencies, a ferrite bead has a high impedance and creates a large ground potential difference between the planes.
- Use multiple decoupling capacitors with different values. Every capacitor has a self-resonant frequency so be careful. Refer to Sect. 13.1 in Power Integrity for more information.
- For PC board stackup, add as many power and ground planes as possible. Keep the power and ground planes next to each other to ensure low impedance stackup or large natural capacitance stackup.
- Add an EMI pi filter on all the signals exiting the box or entering the box.
- If the system fails EMI tests, find the source by tracing the failed frequencies to their source. For example, assume the design fails at 300 MHz but there is nothing on the board running at that frequency. The source is likely a third harmonic of a 100 HMz signal.
- Determine if the failed frequencies are common mode or differential mode. Remove all the cables connected to the box. If the radiation changes, it is common mode, if not, then it is differential mode. Then, go to the clock source and use termination or decoupling techniques to reduce the radiation. If it is common mode, add pi filters to the inputs and outputs. Adding a common choke onto the cable is an effective but expensive method of reducing EMI.

References

1. Federal Communication Commission, Unintentional Radiators, Title 47 (47CFR), Part 15 B (2005), http://www.fcc.gov/oet/info/rules/part15/part15-91905.pdf
2. O. Henry, *Noise Reduction Techniques in Electronic Systems* (Prentice-Hall, Hoboken, NJ, 1988)
3. J. Renolds, *DDR PCB Routing Tutorial* (Texas Instruments Inc., Dallas, TX, 2003)
4. H. Johnson, M. Graham, *High-Speed Signal Propagation* (Prentice-Hall, Hoboken, NJ, 2003)
5. M. Mark, *Printed Circuit Board Design Techniques for EMC Compliance* (The Institute of Electrical and Electronics Engineers, New York, 2000)
6. M. Mark, *Analysis on the Effectiveness of Image Planes Within a Printed Circuit Board* (The Institute of Electrical and Electronics Engineers, New York, 1996)

Index

Printed in the United States
by Baker & Taylor Publisher Services